LINEAR PROGRAMMING

Principles and Applications

Second

L. S. Srinath

Professor of Mechanical Sciences
Indian Institute of Science
Bangalore

M

First published in India 1975, 1982
by Affiliated East-West Press Private Limited

First published in the United Kingdom 1983

Published by
THE MACMILLAN PRESS LIMITED
London and Basingstoke
Companies and representatives
throughout the world

Printed in Hong Kong

ISBN 0 333 36222 5

To Nina

Contents

Preface

In the seven years that have elapsed since the publication of the first edition of this book, the subject of linear programming has acquired new dimensions. In bringing out the second edition, therefore, the content of its forerunner had to be revised and additional topics were required to be included. The new edition is up-to-date and more comprehensive and, accordingly, bears a modified title, namely, *Linear Programming Principles and Applications*.

Linear programming deals with problems whose structure is made up of variables having linear relationships with one another. It is used to maximize or minimize a given objective function which is linear and is governed by linear constraints.

As in the case of the first edition, the purpose here too is to introduce the reader to the principles of linear programming, the techniques of its application, and the variety of problem situations where it can be effectively used. The style adopted will suit both management science people and engineers who are increasingly using the techniques of linear programming to solve their practical problems. Beginning with a fairly elementary approach to optimization problems, the discussion grows detailed, drawing illustrative examples and problems from a wide field of activity. The coverage however remains self-contained and straightforward, and no sophisticated mathematical background is needed to follow the content.

A major departure from the earlier edition is the replacement of the Cases by four new chapters. One of these deals with parametric linear programming, another with integer programming, and the remaining two with the theory of games and its aspects associated with linear programming. The inclusion of these topics should increase the usefulness of the book.

The preparation of the manuscript was aided by the Curriculum Development Centre established at the Indian Institute of Science, Bangalore, by the Ministry of Education and Culture; it is a pleasure to acknowledge their financial assistance. Also, I wish to express my appreciation of the careful editing of my manuscript by the editorial team at East-West Press.

Bangalore
January 1982

L. S. Srinath

1

Elementary Problems in Linear Programming

1.1 INTRODUCTION

All organizations, whether large or small, face situations where decision-making is based on a quantitative analysis to achieve a desired goal. The problem, generally, is of allocating the resources in an efficient manner in order to maximize the profit, or minimize the loss, or utilize the production capacity to the maximum extent. Correct decision-making requires that the problem be capable of being stated in terms of mathematical concepts, such as numbers, lines, and functions. When several solutions are available to the organization, each solution must be associated with a figure of merit to enable the management to compare the potential results of the different methods. Further, selection of the most suitable solution requires a criterion for recognition of the superiority of one method over the others.

Problems that deal with the maximization or the minimization of *objective functions* are called optimization problems. In this chapter, we shall discuss elementary optimization problems which involve only up to two variables and introduce the reader to expressions such as *solution space*, *region of feasible solutions*, *optimum solution*, *decision variables*, and *constraint conditions*. As the elementary problems can be solved geometrically, a clear understanding of these expressions is obtained.

Linear programming deals with the optimization of linear functions subject to linear constraints. Such optimization problems are solved by an iterative process known as the *Simplex method* which is due to George B. Dantzig and is one of the finest contributions of the twentieth century to applied mathematics.

1.2 SINGLE-ANSWER PROBLEMS

Before we discuss how an optimization problem arises in practice, let us consider two elementary problems which may be called single-answer problem situations.

Problem 1.1 A manufacturer who has 18 tonnes of a special variety of steel is committed by contract to supply 7.6 tonnes of bolts and nuts made of this material to a certain company. But he becomes reluctant to supply these bolts and nuts, and with reason: first, there is a 5% loss in steel when it is processed into bolts and nuts; second, the production costs have risen since the signing of the contract, but the selling price of bolts and nuts remains the same; and third, he can make more profit by selling the steel as raw material to another firm. What is the amount of steel the manufacturer can sell as raw material while still honouring his contract?

This simple problem with one variable can be solved algebraically. If x is the number of tonnes of steel the manufacturer can afford to sell as raw material, he has $(18 - x)$ tonnes of steel available to produce bolts and nuts. But, out of this, 5% is lost during the manufacturing process. Hence, the actual amount of bolts and nuts produced is

$$(18 - x) - (18 - x)\frac{5}{100} = (18 - x)(1 - 0.05).$$

This must be equal to 7.6 tonnes. Hence, we get

$$(18 - x)(1 - 0.05) = 7.6, \tag{1.1}$$

i.e.,

$$18 - x = 7.6/0.95$$

or

$$x = 10 \text{ tonnes.}$$

Therefore, 10 tonnes is the amount of steel the manufacturer can sell as raw material.

Problem 1.2 A farmer has 40 acres of land, a part of which is categorized as wet and the rest as dry. He decides to grow rice on one half of his wet land and lets the other half lie fallow. He cultivates the whole of his dry land. The total earnings from his wet- and dry-land cultivation, at the rate of \$ 1200 per acre from the wet land and \$ 800 per acre from the dry land, is \$ 27,000. How many acres of wet land and of dry land did the farmer possess?

This elementary problem involves two variables, as compared with Problem 1.1 which has only one variable. If x and y denote respectively the wet- and the dry-land holding, the first equation we get is

$$x + y = 40. \tag{1.2}$$

The second equation based on the farmer's earnings is

$$(x/2)1200 + 800y = 27{,}000$$

or

$$600x + 800y = 27{,}000. \tag{1.3}$$

Equations (1.2) and (1.3) are solved simultaneously, yielding $x = 25$ acres (wet land) and $y = 15$ acres (dry land).

Problems 1.1 and 1.2 do not involve any optimization process. They represent single-answer or single-set-answer problem situations. This means that such problems, if at all they have answers, will have unique answers such as 10 tonnes of steel in Problem 1.1, and 25 acres of wet land, 15 acres of dry land in Problem 1.2. The fact that the number of unknowns is equal to the number of equations enables us to determine the unknowns uniquely.

1.3 MULTI-ANSWER PROBLEMS

Problems 1.1 and 1.2 can, however, be reworded to introduce the concept of optimization. Let us now reword Problem 1.1.

Problem 1.3 A manufacturer who has 18 tonnes of a special variety of steel is committed by contract to supply *at least* 7.6 tonnes of bolts and nuts made of this material to a certain company. But he becomes reluctant to supply these bolts and nuts, and with reason: first, there is a 5% loss in steel when it is processed into bolts and nuts; second, the production costs have risen since the signing of the contract, but the selling price of bolts and nuts remains the same; and third, he can make more profit by selling the steel as raw material to another firm. What is the *maximum amount* of steel the manufacturer can sell as raw material while still honouring his contract?

The difference between Problems 1.1 and 1.3 is obvious. In Problem 1.3, we have introduced the additional words *at least* and *maximum amount*. These words alter the situation in Problem 1.1. If we take into account only the words *at least*, the amount of steel the manufacturer can sell as raw material can vary from 0 tonne to 10 tonnes, since the contract binds him to supply at least 7.6 tonnes of bolts and nuts, and he can use, if he so wishes, all his 18 tonnes of special steel to manufacture bolts and nuts. This comes to 17.1 tonnes of bolts and nuts, as the 5% loss during production has to be taken into account. In this case, he will have no steel to sell as raw material.

But the manufacturer can make more profit by selling the steel as raw material and, the greater the amount he sells, the more will be his profit. The maximum amount he can afford to sell is 10 tonnes (Problem 1.1). Hence, if we ignore the words *maximum amount*, the answer to Problem 1.3 can vary from 0 tonne to 10 tonnes. However, this is not a single-answer problem situation as in Problem 1.1. If the profit is to be *maximized*, the answer, i.e., the *optimum solution*, is 10 tonnes.

The foregoing situation can be represented graphically, as in Fig. 1.1. The length OA represents 18 tonnes, the total amount of steel owned by the manufacturer. The amount of 7.6 tonnes of bolts and nuts, which is the minimum he has promised to supply his client, requires 8 tonnes of raw material. This is represented by OB which is the minimum amount of steel required to produce the bolts and nuts. However, the manufacturer can use any amount up to 18 tonnes, if he so wishes, for the production process. Hence, the *solution space* for the amount of steel he can afford to sell as raw material can vary from A to B, the shaded portion in

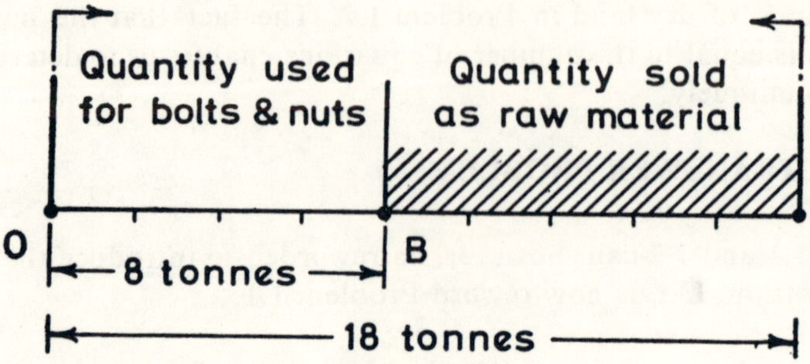

FIGURE 1.1

Fig. 1.1. If the profit is to be *maximized*, the material used for production must be *minimized*. Hence, the *optimum solution* for maximum profit (that is to say, the minimum production of bolts and nuts) is AB, that is, 10 tonnes.

In order to make Problem 1.2 an optimization problem, we shall now reword it.

Problem 1.4 A farmer who has 25 acres of wet land and 15 acres of dry land decides to cultivate a part of his wet land and all his dry land. Since he lacks the necessary initial capital to purchase fertilizers, hire tractors, etc., he has to mortgage a part of his wet land. He can raise \$ 480 per acre through mortgage. His anticipated earnings through cultivation are \$ 1200 per acre of wet land and \$ 800 per acre of dry land. In order to cultivate his lands, he needs an initial capital of \$ 240 per acre of wet land and \$ 200 per acre of dry land. If the farmer wants to *maximize* his earnings, how many acres of wet land should he mortgage?

If we leave out the words *maximize his earnings*, the farmer can mortgage all of his wet-land holdings, or none, or any amount from 0 acre to 25 acres. His income from the cultivation of the remaining land will depend on the amount he can raise through mortgage. In such a case, the solution space will include any number between 0 and 25.

However, if the earnings are to be *maximized*, the problem requires an optimum solution. The data shows that the net earnings per acre through the cultivation of wet land and dry land are respectively \$ 960 (=1200 − 240) and \$ 600 (=800 − 200), and these are higher than the mortgage income of \$ 480 per acre of wet land. Hence, it is more profitable to cultivate the maximum acreage and raise the necessary initial capital by mortgaging the *minimum* amount of wet land. If x is the minimum number of acres of wet land to be mortgaged, the money raised through mortgage is \$ $280x$. This should cover the necessary initial expense of $(25 - x) \times 240$ plus

15×200. Hence, we get

$$(25 - x)240 + (15)(200) = 480x$$

or

$$x = 12.5.$$

After mortgaging 12.5 acres of wet land, the farmer can cultivate the remaining 12.5 acres of wet land as well as 15 acres of dry land, getting a total return of $(12.5 \times 1200) + (15 \times 800) =$ \$ 27,000 (Problem 1.2). The farmer can, of course, choose to mortgage more than 12.5 acres of wet land, at the rate of \$ 480 per acre, and thereby have less area available for cultivation. But, in this case, his profit would be lower, and the solution would not be an optimum solution, as the amount earned from mortgage is much less than that from cultivation.

Problems 1.3 and 1.4 show how an optimization problem might arise in practice.

1.4 GRAPHICAL SOLUTION

Linear programming problems involving two variables can be effectively solved by a graphical approach. We shall explain this technique through the solution of a few specific problems as they actually occur.

Problem 1.5 A firm manufactures two types of screws, Staybrite and Regular. Staybrite screws earn a profit of \$ 3.00 per thousand, and Regular, \$ 1.50 per thousand. Staybrites require a special chemical treatment. If all the available facilities in the firm are utilized, 40,000 Staybrite screws can be produced per day as against 60,000 Regulars per day. The chemicals required for Staybrites are restricted in supply and are sufficient for a maximum of 30,000 screws per day. The total packing capacity in the firm is restricted to 50,000 screws per day. How many Staybrites and how many Regulars should the firm manufacture to get the maximum profit?

Let x and y represent, in thousands, the number of Staybrites and the number of Regulars, respectively, produced per day. Obviously, the variables x and y can be zero or positive, but can never be negative. This is known as the *positivity condition*. The profit on x thousand Staybrites and y thousand Regulars is $3x + 1.5y$. This is called the *objective function* —the object being to *maximize* the profit—which we shall denote by Z. The problem is to

$$\text{maximize } Z = 3x + 1.5y. \tag{1.4}$$

The variables x and y appear in a linear manner in the objective function and cannot assume all the possible positive values. They are restricted by certain conditions, such as plant facility, packing capacity, and chemicals available. These governing conditions are called *constraints* or *constraint conditions*. In our example, if we consider plant facility alone, the maximum value that x can assume, when $y = 0$, is 40. The actual production, however, can be less. Hence, we write this as

$$x \leqslant 40. \tag{1.5a}$$

Similarly, the production of Regulars, based on plant facility alone, when x is zero, is governed by

$$y \leqslant 60. \tag{1.5b}$$

Since the rate of production of Staybrites is 5000 per hour and that of Regulars 7500 per hour, the total time taken to produce x thousand Staybrites and y thousand Regulars is $x/5 + y/7.5$. Assuming that the production department works for 8 hours per day, we have

$$x/5 + y/7.5 \leqslant 8$$

or

$$3x + 2y \leqslant 120. \tag{1.6}$$

This satisfies both conditions, namely, (1.5a) and (1.5b).

The packing capacity of the firm imposes a constraint on the sum of x and y which cannot exceed 50, i.e.,

$$x + y \leqslant 50. \tag{1.7}$$

The constraint imposed by the chemicals gives

$$x \leqslant 30. \tag{1.8}$$

Hence, there are five constraint conditions expressed by (1.5a), (1.5b), (1.6), (1.7), and (1.8). Subject to these, we have to maximize the objective function Z, given by Eq. (1.4). The variables x and y appear linearly in all the constraints.

Let us consider Fig. 1.2, where the horizontal axis represents the number

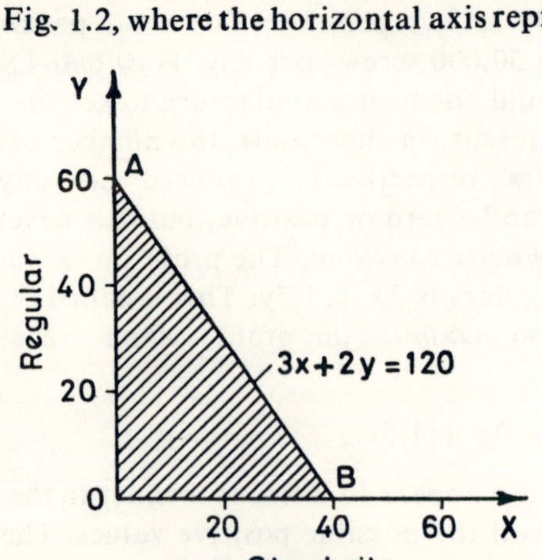

FIGURE 1.2

of Staybrites, and the vertical axis, the number of Regulars, produced in thousands per day. Inequality (1.6), coupled with the fact that $x \geqslant 0$, $y \geqslant 0$, imposes the condition that the values of x and y should lie within the area bounded by the X-axis, the Y-axis, and the straight line AB represented by the equation $3x + 2y = 120$.

Condition (1.7) alone stipulates that the sum of x and y should not exceed 50, i.e., x and y should lie within OC, OD, and CD, as shown in Fig. 1.3. When conditions (1.6) and (1.7) are imposed together, x and y

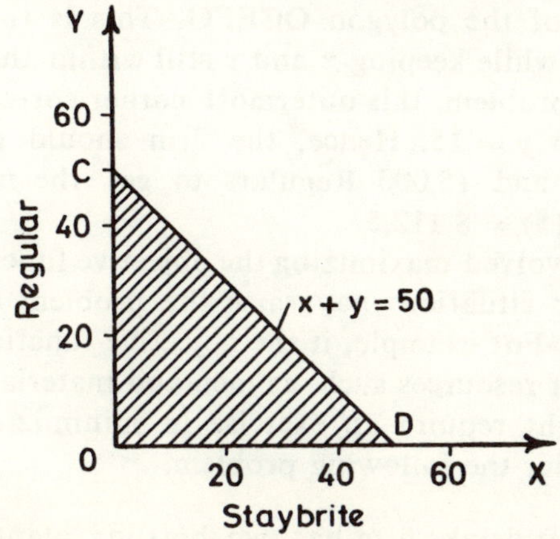

FIGURE 1.3

can lie only in the area common to the regions OABO (Fig. 1.2) and OCDO (Fig. 1.3). When, in addition, condition (1.8) is imposed, the values of x and y have to be in the shaded region OCEFGO (see Fig. 1.4).

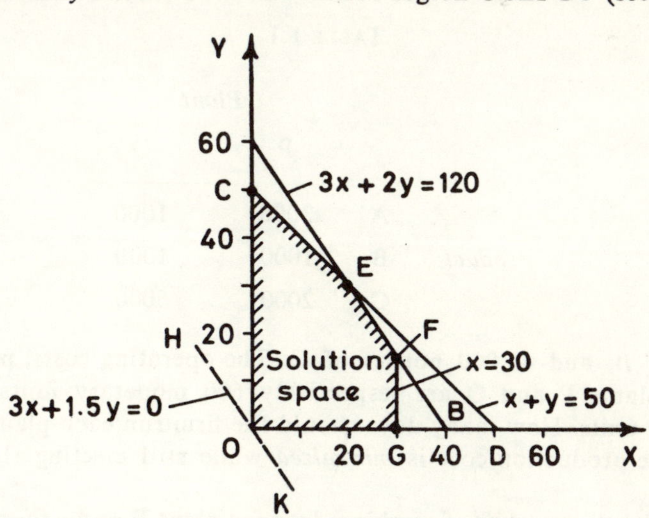

FIGURE 1.4

Polygon OCEFG is the *solution space* for Problem 1.5. This is also sometimes called the region of *feasible solutions*. We have to select the values of x and y lying within the solution space such that the profit, expressed by Eq. (1.4), is maximized. To do this, we notice that when Z is made zero Eq. (1.4) becomes $3x + 1.5 = 0$, and this is represented by the broken line HK passing through O (Fig. 1.4). Its slope, i.e., the tangent of the angle with the X-axis, is $(-3/1.5) = -2$. As the value of Z is increased from zero, the line HK starts moving to the right, parallel to itself. The greater the value that Z can assume, the more will be the firm's profit. Hence, the value of Z can be increased until HK cuts the outermost corner of the polygon OCEFG. This is the maximum value that Z can obtain while keeping x and y still within the region of feasible solutions. In our problem, this outermost corner corresponds to point F, giving $x = 30$ and $y = 15$. Hence, the firm should produce, per day, 30,000 Staybrites and 15,000 Regulars to get the maximum profit of $(3 \times 30) + (1.5 \times 15) = \$\ 112.5$.

Problem 1.5 involved maximizing the objective function. However, the nature of another situation may raise the problem of minimizing the objective function. For example, if the objective function involves losses or consumption of resources such as men and material, then the optimization process might require an exercise in minimizing. To make this clear, let us consider the following problem.

Problem 1.6 A soft drinks firm has two bottling plants, one located at P and the other at Q. Each plant produces three different soft drinks, A, B, and C. The capacities of the two plants, in number of bottles per day, are as given in Table 1.1. A market survey indicates that during the month of April there will be a demand for 24,000 bottles of A, 16,000

TABLE 1.1

		Plant	
		P	Q
	A	3000	1000
Product	B	1000	1000
	C	2000	6000

bottles of B, and 48,000 bottles of C. The operating costs, per day, of running plants P and Q are respectively 600 monetary units and 400 monetary units. How many days should the firm run each plant in April so that the production cost is *minimized* while still meeting the market demand?

Let x be the number of working days for plant P and y the number of working days for plant Q in April. The production cost for April would then be $600x + 400y$. This is our objective function and the problem is to

$$\text{minimize } Z = 600x + 400y. \tag{1.9}$$

During the working days, the firm would produce (number of bottles)

$$3000x + 1000y \text{ of A,}$$

$$1000x + 1000y \text{ of B,}$$

$$2000x + 6000y \text{ of C.}$$

Since these production amounts should meet the market requirement, we have the constraints

$$3000x + 1000y \geqslant 24{,}000, \quad (1.10a)$$

$$1000x + 1000y \geqslant 16{,}000, \quad (1.10b)$$

$$2000x + 6000y \geqslant 48{,}000. \quad (1.10c)$$

The solution space for x and y satisfying these constraints is the shaded region in Fig. 1.5. The objective function, when $Z = 0$, gives the equation

$$600x + 400y = 0 \quad (1.11)$$

which is represented by the broken line passing through O (Fig. 1.5). Note

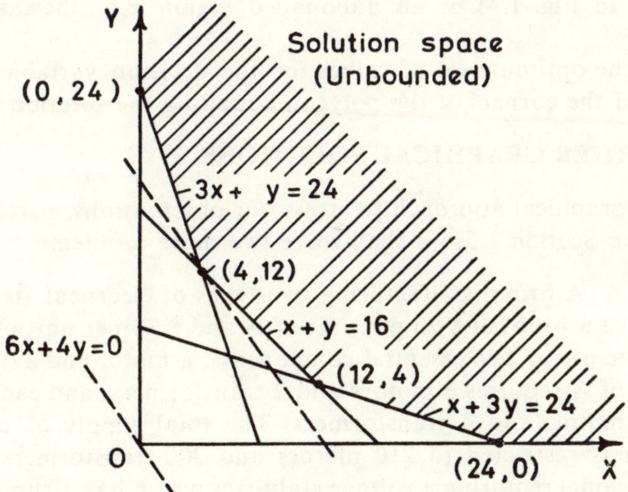

FIGURE 1.5

that this is not a feasible solution. If Z is increased from zero, this broken line moves to the right, parallel to itself. Since we are interested in *minimizing* Z, we increase the value of Z until the broken line touches the *nearest corner* of the shaded region. This is the minimum value Z can obtain while keeping x and y within the region of feasible solutions. The values of x and y corresponding to this corner are respectively 4 and 12. Hence, if the firm operates plant P for 4 days and plant Q for 12 days, the production cost would be a minimum—that is, $(600 \times 4) + (400 \times 12)$

= 7200 monetary units—and the expected demand of the market would also be met.

1.5 SOME OBSERVATIONS

Problems 1.5 and 1.6 are optimization problems and bring the following important points to our attention:

(i) In every optimization problem, there is an objective function which is to be either maximized or minimized.

(ii) The variables that enter into the problem are called decision variables. They cannot assume arbitrary or unrestricted values. They are normally bound by certain conditions called the constraint conditions or merely constraints.

(iii) The variables are non-negative.

(iv) The objective function and the constraint conditions are linear in form with respect to the decision variables. Hence, the problems are classified as *linear programming problems.*

(v) The constraint conditions generally give an area or a region of feasible solutions, also known as the solution space.

(vi) The region of feasible solutions may be a bounded region (e.g., OCEFGO in Fig. 1.4) or an unbounded region (e.g., the shaded region in Fig. 1.5).

(vii) The optimum set of values for the decision variables coincides with one of the corners of the polygon bounding the solution space.

1.6 FURTHER GRAPHICAL SOLUTIONS

Using the graphical approach to stress our observations, particularly (vi) and (vii), in Section 1.5, we shall solve two more problems.

Problem 1.7 A firm manufacturing two types of electrical items, A and B, can make a profit of \$ 20 per unit of A and \$ 30 per unit of B. Both A and B make use of two essential components, a motor and a transformer. Each unit of A requires 3 motors and 2 transformers, and each unit of B requires 2 motors and 4 transformers. The total supply of components per month is restricted to 210 motors and 300 transformers. Type B is an export model requiring a voltage stabilizer which has a supply restricted to 65 units per month. How many each of A and B should the firm manufacture per month to maximize its profit? How much is this profit?

Let x and y be the number of units produced per month of type A and type B, respectively. The object is to

maximize $Z = 20x + 30y$

subject to the constraints

$3x + 2y \leqslant 210,$

$2x + 4y \leqslant 300,$

$y \leqslant 65.$

The solution space is the shaded polygon in Fig. 1.6. The corners of the polygon correspond to the values (0, 0); (0, 65); (20, 65); (30, 60); and (70, 0).

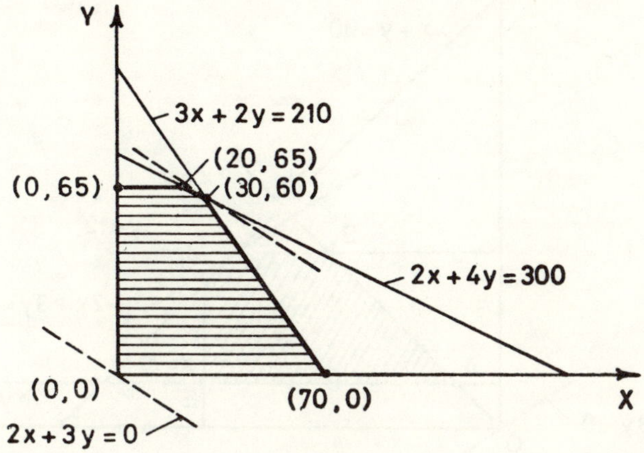

FIGURE 1.6

The optimum solution is 30 units of type A and 60 units of type B, yielding a profit of \$ 2400. This is the *outermost* corner cut by the broken line representing the objective function Z.

Problem 1.8

Determine $x \geqslant 0$, $y \geqslant 0$

so as to

maximize $Z = 2x + 3y$

subject to the constraints

$$x + y \leqslant 30,$$
$$y \geqslant 3,$$
$$y \leqslant 12,$$
$$x - y \geqslant 0,$$
$$x \leqslant 20,$$
$$x \geqslant 0,$$
$$y \geqslant 0.$$

The graphical solution is shown in Fig. 1.7. It is seen that the constraints define a *closed* polygon, ABCDE. This bounded region contains all *feasible solutions*.

The line $2x + 3y = 0$ is shown by the broken line passing through O. Since the problem is to maximize Z, the *outermost* corner of the polygon

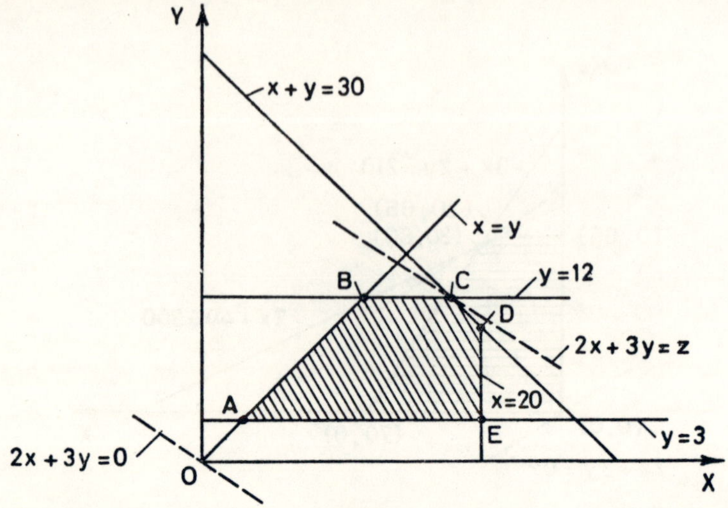

FIGURE 1.7

cut by this line, when it is moved parallel to itself, is C. This point has the values

$$x = 18, \qquad y = 12.$$

The corresponding value of Z is $[(2 \times 18) + (3 \times 12) =]\,72$.

1.7 UNIQUENESS

In the problems discussed so far, the broken line representing the objective function identified a *single corner* which was either closest (in the case of minimization) or farthest (in the case of maximization) from its origin in a direction perpendicular to the objective function line. This provided a *unique* optimal solution.

Situations may arise where there is not one optimal solution but *several optimal solutions*. This occurs when the line representing the objective function is parallel to one of the lines bounding the solution space. For example, consider Problem 1.5. Let the profit of Staybrites be $ 3 per thousand and of Regulars, $ 2 per thousand. The objective function is $Z = 3x + 2y$ instead of $Z = 3x + 1.5y$ (as in Problem 1.5). Now, the objective function line will be parallel to EF and there will not be, as in the original problem, a unique outermost corner cut by the objective function line. All the points from E to F lying on EF will represent optimal solutions, and all will give the same maximum profit of $ 120. Hence, the number of Staybrites and the number of Regulars can vary from (30,000; 15,000) to (20,000; 30,000), the total production per day being controlled by the equation $3x + 2y = 120$.

Similarly, in Problem 1.6, if the operating costs of running plants P and Q are equal at R monetary units per day, then the objective function is $Z = Rx + Ry$. The line representing this will be parallel to the line $x + y = 16$ which bounds the region of feasible solutions. Hence, there will be a series of optimal solutions all satisfying the condition $x + y = 16$ and lying between ($x = 3$, $y = 12$) and ($x = 12$, $y = 4$).

EXERCISES

1. A small manufacturing firm produces two types of gadgets, A and B, which are first processed in the foundry, then sent to the machine shop for finishing. The number of man-hours required in each shop for the production of each unit of A and of B, and the number of man-hours the firm has available per week are as tabulated here. The profit on the

	Foundry	*Machine Shop*
Gadget A	10	5
Gadget B	6	4
Firm's Capacity per week	1000	600

sale of A is \$ 30 per unit as compared with \$ 20 per unit of B. How many each of A and of B should be produced per week in order to maximize the profit? What is the profit obtained in such a case?
[*Ans.* A: 40; B: 100; profit: \$ 3200.]

2. If, in Exercise 1, the profits are \$ 30 and \$ 18 respectively for A and B, how many each of A and of B should be produced per week to maximize the profit and how much is this profit?
[*Ans.* Profit: \$ 3000.]

3. Draw the polygon defined by the inequalities

$$3x_1 - 2x_2 \leqslant 12,$$

$$x_1 - 6x_2 \leqslant 1,$$

$$-x_1 + 2x_2 \leqslant 4,$$

$$x_1 \geqslant 0,$$

$$x_2 \geqslant 0,$$

and identify their extreme-point solutions.
[*Ans.* (0, 0); (0, 2); (8, 6); (35/8, 9/16); (1, 0).]

4. Maximize the linear form

$$Z = -3x_1 + 6x_2$$

subject to the constraints

$$x_1 + 2x_2 + 1 \geqslant 0,$$

$$2x_1 + x_2 - 4 \geqslant 0,$$

$$x_1 - x_2 + 1 \geqslant 0,$$

$$x_1 - 4x_2 + 13 \geqslant 0,$$

$$-4x_1 + x_2 + 23 \geqslant 0.$$

[*Ans.* $Z = 15$, $x_1 = 3$, $x_2 = 4$.]

5. Minimize the linear form

$$Z = -3x_1 + 6x_2$$

subject to the same constraints as in Exercise 4.
[*Ans.* $Z = -33$, $x_1 = 5$, $x_2 = -3$.]

6. A firm manufactures headache pills in two sizes, A and B. Size A contains 2 grains of aspirin, 5 grains of bicarbonate, and 1 grain of codeine; size B contains 1 grain of aspirin, 8 grains of bicarbonate, and 6 grains of codeine. It has been found by users that it requires at least 12 grains of aspirin, 74 grains of bicarbonate, and 24 grains of codeine for providing immediate effect. Determine the *least* number of pills a patient should take to get immediate relief. Determine also the quantity of codeine consumed by the patient.
[*Ans.* Size A: 2 pills; size B: 8 pills; codeine content: 50 grains.]

7. A confectioner manufactures two types of biscuits, Regular and No-Cal. The Regulars sell at a profit of 40 ¢ per box, whereas the No-Cals bring in a profit of 50 ¢ per box. The biscuits are processed in three main operations: blending, cooking, and packaging. The average time taken in minutes for each box, for each of the processing operations, is as tabulated here. The blending equipment is available for a maximum of

	Blending	*Cooking*	*Packaging*
Regular	1 min	5 min	3 min
No-Cal	2 min	4 min	1 min

12 machine hours, the cooking facilities for almost 30 hours, and the packaging equipment for more than 15 hours. Determine how many boxes of each type the confectioner should manufacture in order to maximize his profit. How much is this profit?
[*Ans.* Regular: 120; No-Cal: 300; profit: \$ 198.]

8. International Instruments manufactures two types of automobile speedometers, A and B, which they sell to car dealers at a profit of Rs 20 per unit of A and Rs 10 per unit of B. The number of man-hours required, on an average, in each department of International Instruments for the production of each unit of A and of B, and the number of man-hours the firm has available per week for each production run are as tabulated here.

	Assembly	*Plating and Finishing*	*Testing and Packing*
Speedometer A	15	5	1
Speedometer B	6	4	2
Firm's Capacity	3000	1300	500

How many speedometers each of A and of B should International Instruments produce per week in order to realize the maximum profit from each production run, and how much is this profit?
[*Ans.* A: 140; B: 150; profit: Rs 4300.]

9. In Exercise 8, what would be the optimal production programme—i.e., the number required of each type that will maximize profit—if both A and B are sold to the dealers at the same profit of Rs 10 per unit? How much will this optimal profit be?
[*Ans.* A: 100; B: 200; profit: Rs 3000.]

10. Maximize $2x_1 + x_2$
subject to the constraints

$$x_1 - x_2 \leqslant 1,$$

$$x_1 \leqslant 2,$$

$$x_1 + x_2 \leqslant 3,$$

$$x_1 + x_2 \leqslant 4,$$

x_1 and x_2 being non-negative.
[*Ans.* $x_1 = 2$, $x_2 = 1$.]

11. Maximize $2x_1 + x_2$
subject to the constraints

$$2x_1 - 2x_2 \leqslant 1,$$

$$2x_1 - 3x_2 \leqslant 1,$$

$$2x_1 + x_2 \leqslant 2,$$

x_1 and x_2 being non-negative. Also, determine all the basic-feasible solutions.
[*Ans.* $(\frac{5}{6}, \frac{1}{3})$; $(0, 2)$.]

2

Representation in Standard Form

2.1 INTRODUCTION

In Chapter 1, we considered optimization problems involving only up to two variables. We were able to solve these problems by first determining graphically the solution space, then recognizing the optimum values for the variables. This graphical approach was possible since the two variables could be represented by two axes lying in a plane.

When the number of variables is more than two, a graphical solution to optimization problems may not be possible. A general method for solving these problems is then required. Before we discuss this method, let us consider how optimization problems involving more than two variables, and which can be solved by linear programming techniques, may arise in practice.

Problem 2.1 A tea company sells four brands of blended tea which we shall call brands 1, 2, 3, and 4. Each brand is a mixture in varying proportions of three types of tea leaves, A, B, and C. The amount of each of the three types in each kilogram of the four brands and the selling price per kilogram of each brand are shown in Table 2.1. A tea packing and

TABLE 2.1

Type	*Brand*				*New Brand*
	1	2	3	4	
A	0.2	0.4	0.3	0.6	0.42
B	0.3	0.3	0.4	0.2	0.27
C	0.5	0.3	0.3	0.2	0.31
Selling Price per kg	\$ 2.25	\$ 3.00	\$ 2.50	\$ 3.25	Min *Z*

export firm has discovered that each kilogram of the best blend should contain 0.42 kg of A, 0.27 kg of B, and 0.31 kg of C. How many kilograms of each of the four brands should the export firm buy in order to produce 1 kg of its own new brand while keeping its total purchasing cost at a minimum? How much is this cost?

In this blending problem, let x_1, x_2, x_3, and x_4 be respectively the amounts of brands 1, 2, 3, and 4 in each kilogram of the new brand. All these x's are obviously greater than or equal to zero. The problem can now be stated as:

$$\text{Minimize } Z = 2.25x_1 + 3x_2 + 2.5x_3 + 3.25x_4 \tag{2.1}$$

subject to the constraints

$$0.2x_1 + 0.4x_2 + 0.3x_3 + 0.6x_4 = 0.42, \tag{2.2}$$

$$0.3x_1 + 0.3x_2 + 0.4x_3 + 0.2x_4 = 0.27, \tag{2.3}$$

$$0.5x_1 + 0.3x_2 + 0.3x_3 + 0.2x_4 = 0.31, \tag{2.4}$$

$$x_1 + x_2 + x_3 + x_4 = 1, \tag{2.5}$$

$$x_1, x_2, x_3, x_4, \text{ all} \geqslant 0. \tag{2.6}$$

Constraint (2.5) is not an independent equation, because it can be obtained by the addition of Eqs. (2.2), (2.3), and (2.4). The problem as stated is a problem in linear programming, the objective fuction Z being linear in form and the constraint equations also being linear in the variables x_1, x_2, x_3, and x_4. The constraint conditions define the boundary of a region containing feasible solutions. The basic purpose of linear programming is to select from among all feasible solutions an optimal solution which either maximizes or minimizes (here minimizes) the objective function. Generally, the constraint conditions may be inconsistent or contradictory, in which case the problem may not have a feasible solution. Further, the constraint conditions may not define a closed region and the problem may not have a finite optimal solution. We shall discuss these questions later.

Problem 2.2 A part-time dairy-milk distributor has to supply milk to three areas, A, B, and C. Because of the peculiarities of each area, for instance, one-way streets, long driveways, and difficult access, his delivery wages per house differ according to its location: 3 ¢ in area A, 4 ¢ in area B, and 5 ¢ in area C. From experience, he has found that he has to spend, on an average, 1 minute per house in A, 2 minutes per house in B, and 3 minutes per house in C. He can spare only $2\frac{1}{2}$ hours for this milk distribution, and not more than $1\frac{1}{2}$ hours for areas A and B together. The maximum number of bottles he can carry is 120. In these circumstances, to how many houses in each area should he supply milk to earn the maximum per day? How much is this optimal income?

Let the number of houses to which the distributor supplies milk be x_1 in area A, x_2 in area B, and x_3 in area C. The problem is to

$$\text{maximize } Z = 3x_1 + 4x_2 + 5x_3, \tag{2.7}$$

where $x_1 \geqslant 0$, $x_2 \geqslant 0$, $x_3 \geqslant 0$, subject to the constraints

$$x_1 + x_2 + x_3 \leqslant 120, \tag{2.8}$$

$$x_1 + 2x_2 + 3x_3 \leqslant 150, \tag{2.9}$$

$$x_1 + 2x_2 \leqslant 90. \tag{2.10}$$

The objective function Z and the constraint conditions are linear in the variables x_1, x_2, and x_3.

The difference between this problem and the previous one is in the form of their constraint conditions: whereas constraints (2.2) to (2.4) in Problem 2.1 were in the form of equalities, constraints (2.8) to (2.10) in Problem 2.2 are in the form of inequalities. These inequalities can be reformulated to appear as equalities by introducing what are known as *slack variables*. For example, constraint condition (2.8) can be written as

$$x_1 + x_2 + x_3 + x_4 = 120, \tag{2.11}$$

where x_4 is a positive quantity ($\geqslant 0$) such that when it is added to the sum of x_1, x_2, and x_3 the total sum becomes 120. Depending on the values taken by x_1, x_2, and x_3, the new variable x_4 assumes that value that will bring the total sum to 120. Similarly, by adding the positive variables x_5 and x_6 respectively to constraint conditions (2.9) and (2.10), these inequality constraints can also be changed to constraint equations

$$x_1 + 2x_2 + 3x_3 + x_5 = 150, \tag{2.12}$$

$$x_1 + 2x_2 + x_6 = 90. \tag{2.13}$$

The new variables x_4, x_5, and x_6 are the slack variables.

The objective function Z continues to remain the same, and the problem remains unchanged, since the variables x_1, x_2, and x_3 are to be chosen so as to make the value of Z a maximum and, at the same time, satisfy constraint conditions (2.8), (2.9), and (2.10). Depending upon the values finally assumed by these variables, the slack variables x_4, x_5, and x_6 assume appropriate values to satisfy the equalities given by Eqs. (2.11), (2.12), and (2.13). Hence, the optimization problem as stated in Eq. (2.7) with inequality constraint conditions (2.8) to (2.10) can be reformulated as follows:

Determine $x_1 \geqslant 0$, $x_2 \geqslant 0$, $x_3 \geqslant 0$, $x_4 \geqslant 0$, $x_5 \geqslant 0$, $x_6 \geqslant 0$

so as to

$$\text{maximize } Z = 3x_1 + 4x_2 + 5x_3$$

subject to constraint equations

$$x_1 + x_2 + x_3 + x_4 = 120,$$

$$x_1 + 2x_2 + 3x_3 + x_5 = 150,$$

$$x_1 + 2x_2 + x_6 = 90.$$

2.2 STANDARD FORM

We have observed, in Chapter 1, that a linear programming problem involves the determination of the values of certain variables $x_1, x_2, \ldots, x_n$ such that a linear function of these variables, called the objective function, assumes an optimum value (maximum or minimum) when these variables are subject to certain constraint conditions which are also linear functions of the variables. All such linear programming problems can be expressed in a *standard form* as follows:

$$\text{Determine } x_1 \geqslant 0, x_2 \geqslant 0, \ldots, x_n \geqslant 0 \tag{2.14}$$

so as to

$$\text{minimize } Z = c_1x_1 + c_2x_2 + \ldots + c_nx_n \tag{2.15}$$

subject to the constraint conditions expressed as equalities:

$$\begin{aligned} a_{11}x_1 + a_{12}x_2 + \ldots + a_{1n}x_n &= b_1 \\ a_{21}x_1 + a_{22}x_2 + \ldots + a_{2n}x_n &= b_2 \\ &\vdots \\ a_{m1}x_1 + a_{m2}x_2 + \ldots + a_{mn}x_n &= b_m \end{aligned} \tag{2.16}$$

The constants c_j $(j = 1, 2, \ldots, n)$ in the objective function are called *cost coefficients*; the constants b_i $(i = 1, 2, \ldots, m)$ defining the constraint requirements are called *stipulations*; and the constants a_{ij} $(i = 1, 2, \ldots, m$ and $j = 1, 2, \ldots, n)$ are called *structural coefficients.* The sign conditions imposed by (2.14) are known as the *non-negativity requirements.*

We shall now consider how a linear programming optimization problem arising in practice can be recast in a standard form.

Case (*a*) If a particular problem aims to maximize the objective function, for example,

$$\text{maximize } A = c_1x_1 + c_2x_2 + \ldots + c_nx_n,$$

this can be reformulated into a minimization problem by multiplying the objective function by (-1), since the maximization of a quantity is equivalent to the minimization of the negative of that quantity. Hence, the original objective can be written as:

$$\text{Minimize } (-A) = -c_1x_1 - c_2x_2 - \ldots - c_nx_n.$$

Putting $Z = -A$, we get:

$$\text{Minimize } Z = -c_1x_1 - c_2x_2 - \ldots - c_nx_n.$$

Case (*b*) The constraint conditions as expressed in the standard form are equalities. But, in practice, the constraint conditions generally appear as inequalities, with greater-than or less-than signs, or in a mixed form, some constraints being equalities, and others, inequalities. In order to replace the inequality constraint conditions by constraint *equations*, we introduce slack variables which are positive, as shown in Problem 2.2. Let, in a particular problem, the constraint condition appear as

$$a_{i1}x_1 + a_{i2}x_2 + \ldots + a_{ip}x_p \leqslant b_i.$$

By adding a suitable *positive* quantity, x_{p+1}, to the left-hand side, the two sides can be equated. The value of x_{p+1} would, of course, depend on the values assumed by the other x's in the particular equation. The inequality constraint can therefore be written as

$$a_{i1}x_1 + a_{i2}x_2 + \ldots + a_{ip}x_p + x_{p+1} = b_i.$$

It is important to note that the quantity x_{p+1} is a variable, not a fixed quantity. This is so because we shall be dealing in all our problems with multi-answer problem situations, i.e., there will be a solution space as discussed in Chapter 1. Depending on the particular point we select in the solution space (that is to say, the particular values assigned to the variables $x_1, x_2, \ldots, x_n$, all satisfying the given constraint conditions), the new quantity x_{p+1} assumes that value that will make the constraint condition a constraint equation. Hence, it is called a slack *variable*. When an optimum solution is obtained, the slack variable also takes a specific value.

If the constraint condition appears with a greater-than or equal-to sign, for instance,

$$a_{i1}x_1 + a_{i2}x_2 + \ldots + a_{ip}x_p \geqslant b_i,$$

this also can be changed into an equation by subtracting the positive quantity x_{p+1} from the left-hand side:

$$a_{i1}x_1 + a_{i2}x_2 + \ldots + a_{ip}x_p - x_{p+1} = b_i.$$

Sometimes a distinction is made between a positive slack variable and a negative slack variable. The negative slack variable is then called a *surplus variable*. In our discussion, we shall not make this distinction. The sign of the slack variable will depend on the inequality sign in the constraint.

Case (*c*) So far, we have been assuming that all the variables $x_1, x_2, \ldots, x_n$ are non-negative. However, in actual practice, these variables could be positive or negative. If a variable is negative, it can always be expressed as the difference between two positive quantities. For example, $-4 = (+2) - (+6)$. Hence, a variable x_j can be written as

$$x_j = x_j' - x_j'',$$

where $x_j' \geqslant 0$ and $x_j'' \geqslant 0$. Since x_j' may be greater or less than x_j'', the sign of x_j may be positive or negative. Hence, in any problem, if a variable is *unrestricted* in sign, it can always be expressed as the difference between two non-negative variables, and the problem can once again be recast in the standard form.

To make clearer the points in Cases (a), (b), and (c), let us consider a few problems.

Problem 2.3 Reduce the following linear programming problem to the standard form:

Determine $x_1 \geqslant 0$, $x_2 \geqslant 0$, $x_3 \geqslant 0$

so as to

maximize $A = 2x_1 + x_2 + 4x_3$

subject to the constraints

$$-2x_1 + 4x_2 \leqslant 4,$$

$$x_1 + 2x_2 + x_3 \geqslant 5,$$

$$2x_1 + 3x_3 \leqslant 2.$$

This problem can be reduced to the standard form by converting the inequality constraints into equality forms through the introduction of slack variables. Corresponding to the three inequality constraints, there will be three slack variables, x_4, x_5, and x_6. The problem becomes:

Determine $x_j \geqslant 0 \qquad (j = 1, 2, 3, 4, 5, 6)$

so as to

minimize $Z = (-A) = -2x_1 - x_2 - 4x_3$

subject to the constraints

$$-2x_1 + 4x_2 + x_4 = 4,$$

$$x_1 + 2x_2 + x_3 - x_5 = 5,$$

$$2x_1 + 3x_3 + x_6 = 2.$$

Problem 2.4 Reduce Problem 2.3 to the standard form, where x_1 and x_2 are non-negative, but x_3 is unrestricted.

Let $x_3 = x_3 - x_3''$, where both x_3' and x_3'' are non-negative. Thus, the standard form of the problem will be:

Determine $x_1 \geqslant 0$, $x_2 \geqslant 0$, $x_3' \geqslant 0$, $x_3'' \geqslant 0$, $x_4 \geqslant 0$, $x_5 \geqslant 0$, $x_6 \geqslant 0$

so as to

minimize $Z = -2x_1 - x_2 - 4x_3' + 4x_4''$

subject to the constraints

$$-2x_1 + 4x_2 + x_4 = 4,$$

$$x_1 + 2x_2 + x_3' - x_3'' - x_5 = 5,$$

$$2x_1 + 3x_3' - 3x_3'' + x_6 = 2.$$

Problem 2.5 Put the following problem in the standard form:

$$\text{Maximize } Z = 2x_1 - x_2 + x_3$$

subject to the constraints

$$x_1 + 3x_2 - x_3 \leqslant 20,$$

$$2x_1 - x_2 + x_3 \leqslant 12,$$

$$x_1 - 4x_2 - 4x_3 \geqslant 2,$$

$$x_1 \geqslant 0.$$

We observe that this is a maximization problem and that only x_1 is restricted to be non-negative.

Let $x_1 = y_1$, $x_2 = y_2 - y_3$, $x_3 = y_4 - y_5$, where y_1, y_2, y_3, y_4, and y_5 are all non-negative, i.e., $\geqslant 0$. The problem now becomes:

$$\text{Maximize } Z^* = 2y_1 - y_2 + y_3 + y_4 - y_5$$

subject to the constraints

$$y_1 + 3y_2 - 3y_3 - y_4 + y_5 \leqslant 20,$$

$$2y_1 - y_2 + y_3 + y_4 - y_5 \leqslant 12,$$

$$y_1 - 4y_2 + 4y_3 - 4y_4 - 4y_5 \geqslant 2,$$

$$y_i \geqslant 0 \qquad (i = 1, 2, 3, 4, 5).$$

There are now three inequalities. Hence, introducing three slack variables y_6, y_7, and y_8, all positive, and converting the maximization problem into a minimization one, we get:

$$\text{Minimize } R = (-Z^*) = -2y_1 + y_2 - y_3 - y_4 + y_5$$

subject to

$$y_1 + 3y_2 - 3y_3 - y_4 + y_5 + y_6 = 20,$$

$$2y_1 - y_2 + y_3 + y_4 - y_5 + y_7 = 12,$$

$$y_1 - 4y_2 + 4y_3 - 4y_4 + 4y_5 - y_8 = 2,$$

$$y_1, y_2, \ldots, y_8 \geqslant 0.$$

Instead of making $x_2 = y_2 - y_3$ and $x_3 = y_4 - y_5$, it is possible to write $x_2 = y_2 - y_3$ and $x_3 = y_2 - y_4$, thereby reducing the number of variables by one.

2.3 NON-NEGATIVITY CONSTRAINT

We have shown how when one or more decision variables are unrestricted in sign in an optimization problem, this can be transformed into another optimization problem, involving only non-negative decision variables (the slack variables being always non-negative). The question that arises is whether an optimal solution to the new problem is also an optimal solution to the original problem. It can very easily be shown that this is indeed so. To make this clear, let us consider Problem 2.6.

Problem 2.6

$$\text{Minimize } Z = 3x_1 + x_2$$

subject to

$$x_1 - 3x_2 - x_3 = -3,$$

$$2x_1 + 3x_2 - x_4 = -6,$$

$$2x_1 + x_2 + x_5 = 8,$$

$$4x_1 - x_2 + x_6 = 16,$$

$$x_3, x_4, x_5, x_6, \text{ all } \geqslant 0.$$

(Note that we have already included the slack variables to give constraint equations.)

In this problem, x_1 and x_2 which are the decision variables are unrestricted in sign. Putting $x_1 = y_1 - y_2$, $x_2 = y_3 - y_4$, and $x_3 = y_5$, $x_4 = y_6$, $x_5 = y_7$, and $x_6 = y_8$, we get the new problem:

$$\text{Minimize } Z^* = 3y_1 - 3y_2 + y_3 - y_4$$

subject to

$$y_1 - y_2 - 3y_3 + 3y_4 - y_5 = -3,$$

$$2y_1 - 2y_2 + 3y_3 - 3y_4 - y_6 = -6,$$

$$2y_1 - 2y_2 + y_3 - y_4 + y_7 = 8,$$

$$4y_1 - 4y_2 - y_3 + y_4 + y_8 = 16,$$

$$y_1, y_2, \ldots, y_8 \geqslant 0.$$

When we put $x_1 = y_1 - y_2$, we see that for a given x_1 there are an infinite number of (y_1, y_2) combinations satisfying the equation. For example, $4 = (6 - 2) = (8 - 4), \ldots$ or $-2 = (4 - 6) = (8 - 10), \ldots$. But when y_1 and y_2 are given, there is only one value for x_1. Hence, when we find an optimal solution for the new problem giving specific values for $y_1, y_2, \ldots$, the corresponding unique values for $x_1, x_2, \ldots$ also give an optimal solution for the original problem. If this were not so, there would be another set of values $x_1', x_2', \ldots$ which would make $Z(x_1', x_2', \ldots) < Z(x_1, x_2, \ldots)$. These values of $x_1, x_2, \ldots$ would give a new set of values $y_1', y_2', \ldots$ which would make $Z^*(y_1', y_2, \ldots) = Z(x_1', x_2', \ldots) < Z^*(y_1', y_2, \ldots)$, contradicting the optimality of $(y_1, y_2, \ldots)$. Hence, the solution of the new problem will yield the solution of the original problem.

EXERCISES

1. Put each of the following problems in the standard form:

(a) Minimize $A = 4x_1 + 3x_2 + 3x_3 - 4x_4$

subject to the constraints

$$3x_1 + 8x_2 + x_4 \leqslant 120,$$

$$5x_1 + 2x_2 + x_3 + 2x_4 \geqslant 200,$$

$$x_2 + 2x_3 \geqslant 135,$$

$$x_1 \geqslant 0,\ x_2 \geqslant 0,\ x_3 \geqslant 0,\ x_4 \geqslant 0.$$

(b) Minimize $B = c_1x_1 + c_2x_2 + c_3x_3 + c_4x_4$

subject to the constraints

$$a_{11}x_1 + a_{12}x_2 + a_{14}x_4 = b_1,$$

$$a_{21}x_1 + a_{22}x_2 + a_{23}x_3 \geqslant b_2,$$

$$a_{32}x_2 + a_{33}x_3 \leqslant b_3,$$

$$x_1, x_2, x_3, x_4, \text{ all } \geqslant 0.$$

2. In Exercises 1a and 1b, identify the slack variables and list them either as "slack" or "surplus".
[*Ans.* (a) x_1': slack, x_2': surplus, x_3': surplus; (b) $x_1' = 0$, x_2': surplus, x_3': slack.]

3

Gauss-Jordan Elimination Process

3.1 LINEAR SIMULTANEOUS EQUATIONS

A linear programming problem represented in the standard form includes slack variables in addition to the decision variables of the original optimization problem. Both the decision variables and the slack variables are called *admissible variables* and are treated in the same manner in the process of finding a solution to the problem. Keeping in mind that in the standard form the constraints are in the form of equations, we observe that if there are n admissible variables and an equal number ($m = n$) of constraint equations, all linearly independent and consistent (i.e., if the equations do not contradict one another), then a unique solution which is also the optimal solution exists for these variables.

On the other hand, if the constraint equations exceed the number of admissible variables (i.e., $m > n$), there can be no solution unless $(m - n)$ equations are linearly dependent on the n remaining equations, or redundant. Therefore, our primary interest is in those optimization problems where $m < n$, i.e., where the number of constraint equations is less than the number of admissible variables.

In this chapter, we shall briefly discuss certain aspects of linear simultaneous equations that are of particular interest to us. Let there be m linear equations (which are, actually, the m constraint equations of our optimization problem) involving n variables and let $m < n$:

$$\begin{aligned} a_{11}x_1 + a_{12}x_2 + \ldots + a_{1n}x_n &= b_1 \\ a_{21}x_1 + a_{22}x_2 + \ldots + a_{2n}x_n &= b_2 \\ &\vdots \\ a_{m1}x_1 + a_{m2}x_2 + \ldots + a_{mn}x_n &= b_m \end{aligned} \qquad (3.1)$$

We can also write Eqs. (3.1) in an abridged form as

$$\sum_{j=1}^{n} a_{ij}x_j = b_i \qquad (i = 1, 2, \ldots, m). \qquad (3.2)$$

Let us assume that (i) these m equations are consistent (i.e., the equations as expressed do not contradict one another) and (ii) they are linearly independent. Linear independence means that none of the equations in

(3.1) can be obtained from linear combinations of the other equations. On the other hand, let us consider the equations

$$2x_1 + 4x_2 - 3x_3 + x_4 = 15, \tag{3.3a}$$

$$4x_1 + 3x_2 - 2x_4 = 20, \tag{3.3b}$$

$$2x_1 - 6x_2 + 9x_3 - 7x_4 = -5. \tag{3.3c}$$

These three equations are not linearly independent, since Eq. (3.3c) can be obtained from a linear combination of Eqs. (3.3a) and (3.3b):

$$\text{Eq. (3.3c)} = 2 \cdot \text{Eq. (3.3b)} - 3 \cdot \text{Eq. (3.3a)}.$$

To those equations that are linearly independent and consistent, Statements 1 and 2 apply.

Statement 1 Multiplying any of the equations in (3.1) by a constant other than zero does not change the solution to the set of equations, i.e., if we multiply the first equation by λ_1, the second by λ_2, and so on, we get

$$\begin{aligned}
\lambda_1 a_{11} x_1 + \lambda_1 a_{12} x_2 + \ldots + \lambda_1 a_{1n} x_n &= \lambda_1 b_1 \\
\lambda_2 a_{21} x_1 + \lambda_2 a_{22} x_2 + \ldots + \lambda_2 a_{2n} x_n &= \lambda_2 b_2 \\
&\vdots \\
\lambda_m a_{m1} x_1 + \lambda_m a_{m2} x_2 + \ldots + \lambda_m a_{mn} x_n &= \lambda_m b_m
\end{aligned}$$

Then the solution set for these equations is identical to that for the original ones, i.e., Eqs. (3.1).

Statement 2 If the *i*-th equation is multiplied by λ_i, and the *j*-th by λ_j ($\lambda_i, \lambda_j \neq 0$), and if the sum of these two is used in place of the original *i*-th equation, the solution set remains unaltered, i.e., for the following set of equations the solution set is the same as that for Eqs. (3.1):

$$\begin{aligned}
a_{11} x_1 + a_{12} x_2 + \ldots + a_{1n} x_n &= b_1 \\
a_{21} x_1 + a_{22} x_2 + \ldots + a_{2n} x_n &= b_2 \\
&\vdots \\
(\lambda_i a_{i1} + \lambda_j a_{j1}) x_1 + (\lambda_i a_{i2} + \lambda_j a_{j2}) x_2 + \ldots &= \lambda_i b_i + \lambda_j b_j \\
a_{j1} x_1 + a_{j2} x_2 + \ldots + a_{jn} x_n &= b_j \\
&\vdots \\
a_{m1} x_1 + a_{m2} x_2 + \ldots + a_{mn} x_n &= b_m
\end{aligned}$$

Using Statements 1 and 2, we shall now introduce the *Gauss-Jordan complete elimination procedure* for solving systems of simultaneous linear equations. Once this procedure is understood, it is a fairly simple transition to the Simplex method of solving linear programming problems.

3.2 GAUSS-JORDAN COMPLETE ELIMINATION PROCEDURE

We shall introduce the Gauss-Jordan procedure by means of a specific problem, but it will be evident that the method itself is quite general.

Problem 3.1 From the following three equations in five unknowns, find solutions for x_3, x_4, and x_5 in terms of x_1 and x_2:

$$2x_1 - x_2 + 2x_3 - x_4 + 3x_5 = 14,$$

$$x_1 + 2x_2 + 3x_3 + x_4 = 5,$$

$$x_1 - 2x_3 - 2x_5 = -10.$$

Let us rewrite the given equations by detaching the coefficients to obtain a matrix representation as in Tableau I. Our aim is to eliminate x_3

TABLEAU I

x_1	x_2	x_3	x_4	x_5	
2	−1	***2***	−1	3	14
1	2	3	1	0	5
1	0	−2	0	−2	−10

from all the equations except the first one, x_4 from all the equations except the second, and x_5 from all the equations except the third so that the final set of equations has unit coefficients in place of variables. This is achieved by repeatedly applying the procedures in Statements 1 and 2 (Section 3.1).

First, we eliminate x_3 from the second and third equations by following three steps: (i) We divide each entry in the first row by 2, the coefficient of x_3 in the first equation (this is equivalent to multiplying by 1/2, as explained in Statement 1). Coefficient 2 is called a *pivot* and is shown in ***bold italic*** in the first matrix, as it operates to yield a new first row (see Tableau II) where the coefficient of x_3 is unity. According to Statement 1,

TABLEAU II

x_1	x_2	x_3	x_4	x_5	
1	−1/2	1	−1/2	3/2	7
1	2	3	1	0	5
1	0	−2	0	−2	−10

the solution set for these equations is identical to that for the original ones. (ii) We multiply the new first row by (−3) and add it to the second row. (iii) We multiply the new first row by 2 and add it to the third row.

These three steps result in a new matrix (see Tableau III) in which the coefficient of x_3 in the second and third rows is zero. According to

TABLEAU III

x_1	x_2	x_3	x_4	x_5	
1	−1/2	1	−1/2	3/2	7
−2	7/2	0	5/2	−9/2	−16
3	−1	0	−1	1	4

Statement 2, the solution set for this is identical to that for the original set of equations.

Next, we (i) divide each entry in the second row by the coefficient of x_4 in that row so as to make the coefficient of x_4 unity as in Tableau IV.

TABLEAU IV

x_1	x_2	x_3	x_4	x_5	
−4/5	7/5	0	1	−9/5	−32/5 (second row)

In order to eliminate x_4 from the first and third rows, we (ii) multiply the new second row by 1/2 and add it to the first row; and (iii) multiply the second row by 1 and add it to the third row. This second set of three steps results in the matrix given in Tableau V.

TABLEAU V

x_1	x_2	x_3	x_4	x_5	
3/5	1/5	1	0	3/5	19/5
−4/5	7/5	0	1	−9/5	−32/5
11/5	2/5	0	0	−4/5	−12/5

The third and final iteration eliminates x_5 from the first and second rows. We again achieve this in three steps: (i) divide the entries in the third row by (−4/5); (ii) multiply the new third row by 9/5 and add it to the second row; and (iii) multiply the new third row by (−3/5) and add it to the first row. This final set of three steps successively makes (i) the coefficient of x_5 in the third row unity; (ii) the coefficient of x_5 in the second row zero; and (iii) the coefficient of x_5 in the first row zero. The final matrix has the form shown in Tableau VI.

TABLEAU VI

x_1	x_2	x_3	x_4	x_5	
9/4	1/2	1	0	0	2
−23/4	1/2	0	1	0	−1
−11/4	−1/2	0	0	1	3

Rewriting the final matrix as equations, we get

$$(9/4)x_1 + (1/2)x_2 + x_3 = 2,$$

$$(-23/4)x_1 + (1/2)x_2 + x_4 = -1,$$

$$(-11/4)x_1 - (1/2)x_2 + x_5 = 3.$$

These new equations are obtained from repeated applications of Statements 1 and 2. Hence, the solution set for these equations is identical to that for the original set of equations.

The solutions for x_3, x_4, and x_5 are readily obtained from these new equations as

$$x_3 = 2 - (9/4)x_1 - (1/2)x_2,$$

$$x_4 = -1 + (23/4)x_1 - (1/2)x_2,$$

$$x_5 = 3 + (11/4)x_1 + (1/2)x_2.$$

The important fact to observe in the iterative steps is that, in the final matrix form, each of the variables x_3, x_4, and x_5 (whose values we attempted to find in terms of the remaining variables) appears in one equation only with its coefficient unity. When a given set of equations is so rewritten, it is said to have been expressed in a *canonical form*. Further, the variables x_3, x_4, and x_5 (which were arbitrarily chosen) are called the *basic variables*. There are infinite solutions for x_3, x_4, and x_5, depending on the values given to x_1 and x_2. The solution for these basic variables, when the values of the non-basic variables (x_1 and x_2) are zero, is called the *basic solution* The basic solution of Problem 3.1 is therefore

$$x_3 = 2, \qquad x_4 = -1, \qquad x_5 = 3.$$

We could have, of course, chosen any other variables, for instance, x_1, x_2, and x_3, as our basic variables, and obtained the corresponding basic solution by making the non-basic variables x_4 and x_5 zero. The following example makes this point clear. Let

$$2x_1 + x_2 + x_3 + x_5 = 8,$$

$$3x_1 + 2x_3 + x_4 + 2x_5 = 9,$$

$$x_1 + 3x_5 + x_6 = 5.$$

This set of equations is canonical in form with respect to the variables x_2,

x_4, and x_6. If these are taken as the basic variables, the basic solution is obtained by making the non-basic variables x_1, x_3, and x_5 zero. Hence, in this case, the basic solution is

$$x_2 = 8, \qquad x_4 = 9, \qquad x_6 = 5.$$

Let us consider one more problem where the Gauss-Jordan elimination process is used to solve a given set of linear equations.

Problem 3.2 For the following four equations in six unknowns, choose x_1, x_2, x_3, and x_4 as basic variables and determine the corresponding basic solution:

$$2x_1 + 4x_2 + x_3 + 3x_4 + x_5 + 3x_6 = 20,$$

$$x_1 + 3x_2 - 2x_3 + 2x_5 = 9,$$

$$3x_1 - 2x_2 - x_3 + x_5 - x_6 = -1,$$

$$2x_2 + 3x_3 + 2x_4 + 2x_6 = 11.$$

In order to determine the basic solution, we must first express these equations in a canonical form with respect to the variables x_1, x_2, x_3, and x_4. We write the set of equations in a matrix form with detached coefficients as in Tableau I.

TABLEAU I

x_1	x_2	x_3	x_4	x_5	x_6	
2	4	1	3		3	20
1	3	-2	0	2	0	9
3	-2	-1	0	1	-1	-1
0	2	3	2	0	2	11

First, we divide each entry in the first row by 2, the pivot, to make the coefficient of x_1 unity in that row as in Tableau II. Next, we perform the

TABLEAU II

x_1	x_2	x_3	x_4	x_5	x_6	
1	2	1/2	3/2	1/2	3/2	10
1	3	-2	0	2	0	9
3	-2	-1	0	1	-1	-1
0	2	3	2	0	2	11

following operations separately: (i) multiply the first row by (-1) and add it to the second row; (ii) multiply the first row by (-3) and add it to the third row. The fourth row already has a zero coefficient for x_1. The new matrix then takes the form shown in Tableau III.

TABLEAU III

x_1	x_2	x_3	x_4	x_5	x_6	
1	2	1/2	3/2	1/2	3/2	10
0	*1*	−5/2	−3/2	3/2	−3/2	−1
0	−8	−5/2	−9/2	−3/2	−11/2	−31
0	2	3	2	0	2	11

The coefficient of x_2 in the second row is already unity. Hence, we now perform the following operations separately: (i) multiply the second row by (−2) and add it to the first row; (ii) multiply the second row by 8 and add it to the third row; (iii) multiply the second row by (−2) and add it to the fourth row. After these operations, the new matrix appears as in Tableau IV.

TABLEAU IV

x_1	x_2	x_3	x_4	x_5	x_6	
1	0	11/2	9/2	−1	3	12
0	1	−5/2	−3/2	3/2	−3/2	−1
0	0	*−45/2*	−33/2	21/2	−35/2	−39
0	0	8	5	−3	7/2	13

Further, dividing the third row by (−45/2), the new pivot, we make the coefficient of x_3 in that row equal to unity. The resulting matrix is as given in Tableau V.

TABLEAU V

x_1	x_2	x_3	x_4	x_5	x_6	
1	0	11/2	9/2	−1	3	12
0	1	−5/2	−3/2	3/2	−3/2	−1
0	0	1	11/15	−7/15	7/9	26/15
0	0	8	5	−3	7/2	13

The fourth set of iterations involves the following separate operations: (i) multiply the third row by (−11/2) and add it to the first row; (ii) multiply the third row by 5/2 and add it to the second row;

(iii) multiply the third row by (−8) and add it to the fourth row. These operations yield the matrix given in Tableau VI.

TABLEAU VI

x_1	x_2	x_3	x_4	x_5	x_6	
1	0	0	7/15	47/30	−23/18	37/17
0	1	0	1/3	1/3	4/9	10/3
0	0	1	11/15	−7/15	7/9	26/15
0	0	0	*−13/15*	11/15	−49/18	−13/15

The final iteration involves, first, dividing the fourth row by (−15/13) so as to make the coefficient of x_4 in that row equal to unity, and then performing the following operations separately: (i) multiply the fourth row by (−7/15) and add it to the first row; (ii) multiply the fourth row by (−1/3) and add it to the second row; (iii) multiply the fourth row by (−11/15) and add it to the third row. The final matrix will be as in Tableau VII. The matrix of the detached coefficients is now in a canonical

TABLEAU VII

x_1	x_2	x_3	x_4	x_5	x_6	
1	0	0	0	51/26	−22/9	2
0	1	0	0	8/13		3
0	0	1	0			1
0	0	0	1	−11/13	245/98	1

form with respect to x_1, x_2, x_3, and x_4. Hence, we write the basic solution as

$$x_1 = 2, \qquad x_2 = 3, \qquad x_3 = 1, \qquad x_4 = 1.$$

We observe that, in the process of obtaining the canonical form with respect to x_1, x_2, x_3, and x_4 in Problem 3.2, it was not necessary to work out the operations on the coefficients of x_5 and x_6 at each step, since we intended to finally equate them to zero. Though this observation is true with reference to the operations in Problem 3.2, it does not apply to the Simplex method of solving linear programming problems, where we need the coefficients of the non-basic variables too. Hence, we have also listed the transformations that the coefficients of the non-basic variables undergo during all the operations in Problem 3.2.

3.3 GENERAL PROCEDURE AND TEST FOR INCONSISTENCY

In order to make it clear that the Gauss-Jordan procedure followed in

Problems 3.1 and 3.2 is quite general, we list the steps that reduce a given set of equations to a canonical form. Further, if the given set of equations is *inconsistent* or *incompatible*, this also will be revealed.

Let the system of linear equations be

$$\begin{aligned} a_{11}x_1 + a_{12}x_2 + \ldots + a_{1n}x_n &= b_1 \\ a_{21}x_1 + a_{22}x_2 + \ldots + a_{2n}x_n &= b_2 \\ &\vdots \\ a_{m1}x_1 + a_{m2}x_2 + \ldots + a_{mn}x_n &= b_m \end{aligned} \tag{3.4}$$

We assume, as in Section 3.1, that the number of decision variables $x_1, x_2, \ldots, x_n$ exceeds the number of equations, i.e., $m < n$. Let $x_1, x_2, \ldots, x_m$ be selected as the basic variables. Under the assumption that Eqs. (3.4) do indeed possess a solution, we proceed as follows.

Step 1 Suppose $a_{11} \neq 0$ (otherwise, we shift the variables or equations to make it so). We divide both sides of the first equation in (3.4) by a_{11} so that it takes the form

$$x_1 + a'_{12}x_2 + a'_{13}x_3 + \ldots + a'_{1m}x_m = b'_1. \tag{3.5}$$

Multiplying both sides of Eq. (3.5) successively by $-a_{21}, -a_{31}, -a_{41}, \ldots, -a_{m1}$, and adding the respective resultant equations to the second, third, fourth, ..., m-th equation in (3.4), we reduce Eqs. (3.4) to the form

$$\begin{aligned} x_1 + a'_{12}x_2 + a'_{13}x_3 + \ldots + a'_{1n}x_n &= b'_1 \\ a'_{22}x_2 + a'_{23}x_3 + \ldots + a'_{2n}x_n &= b'_2 \\ a'_{32}x_2 + a'_{33}x_3 + \ldots + a'_{3n}x_n &= b'_3 \\ &\vdots \\ a'_{m2}x_2 + a'_{m3}x_3 + \ldots + a'_{mn}x_n &= b_m \end{aligned} \tag{3.6}$$

Step 2 Suppose $a'_{22} \neq 0$ (otherwise, we shift the variables or equations to make it so). We divide both sides of the second equation in (3.6) by a'_{22} so that it takes the form

$$x_2 + a''_{23}x_3 + \ldots + a''_{2n}x_n = b''_2. \tag{3.7}$$

Multiplying both sides of Eq. (3.7) successively by $-a'_{12}, -a'_{32}, -a'_{42}, \ldots, -a'_{m2}$, and adding the respective resultant equations to the first, third, fourth, ..., m-th equation, as in Step 1, we get

$$\begin{aligned} x_1 \quad + a''_{13}x_3 + \ldots + a''_{1n}x_n &= b''_1 \\ x_2 + a''_{23}x_3 + \ldots + a''_{2n}x_n &= b''_2 \\ a''_{33}x_3 + \ldots + a''_{3n}x_n &= b''_3 \\ &\vdots \\ a''_{m3}x_3 + \ldots + a''_{mn}x_n &= b''_m \end{aligned} \tag{3.8}$$

Remaining Steps We repeat Steps 1 and 2 m times. If the given set of equations is consistent, each of the variables $x_1, x_2, \ldots, x_m$ will appear respectively in one equation only, with their coefficients unity, i.e., they will appear in a canonical form.

Redundancy If r ($<m$) of the m equations in (3.4) are *redundant*, i.e., are not independent but consistent, these r equations will appear in the final canonical form with all the coefficients on the left-hand side equal to zero and their right-hand sides also will be zero.

Inconsistency If Eqs. (3.4) are *inconsistent*, one or more of the equations will have zero on the left-hand side and a non-zero quantity on the right-hand side, such as $0 = b_r^*$, where $b_r^* \neq 0$, indicating that a solution to the given set of equations does not exist.

Problem 3.3 Test the following three equations involving four unknowns for redundancy and/or inconsistency:

$$x_1 + 2x_2 - x_3 - 2x_4 = 1,$$

$$2x_1 + x_2 + x_3 - x_4 = 8,$$

$$x_1 + 3x_2 - 2x_3 - 3x_4 = -1.$$

Table 3.1 lists the iterative steps in the Gauss-Jordan reduction process.

TABLE 3.1

x_1	x_2	x_3	x_4	
1	2	−1	−2	1
2	1	1	−1	4
1	3	−2	−3	−3
1	2	−1	−2	1
0	−3	3	3	6
0	1	−1	−1	−2
1	2	−1	−2	−1
0	1	−1	−1	−2
0	1	−1	−1	−2
1	0	1	0	1
0	1	−1	−1	−2
0	0	0	0	0

Hence, the third equation is redundant. Indeed, it is obtained by subtracting one-third of the second equation from five-thirds of the first.

Problem 3.4 Test the following equations for inconsistency:

$$x_1 + 2x_2 - x_3 - 2x_4 = 1,$$

$$2x_1 + x_2 + x_3 - x_4 = 4,$$

$$x_1 + 3x_2 - 2x_3 - 3x_4 = -5.$$

The table of entries will be identical to Table 3.1 except for the third row. It can be easily checked that the last entry for the third row in Problem 3.4 will give $0 = 2$, indicating that the equations are inconsistent and therefore do not have a solution.

EXERCISES

1. Use the Gauss-Jordan elimination procedure to solve the following equations:

(a)

$$2x_1 + 2x_2 - x_3 + x_4 = 4,$$

$$4x_1 + 3x_2 - x_3 + 2x_4 = 6,$$

$$8x_1 + 5x_2 - 3x_3 + 4x_4 = 12,$$

$$3x_1 + 3x_2 - 2x_3 + 2x_4 = 6.$$

[*Ans.* $x_1 = 1, x_2 = 1, x_3 = -1, x_4 = -1$.]

(b)

$$2x_1 + x_2 + 4x_3 = 4,$$

$$x_1 - 3x_2 - x_3 = -5,$$

$$3x_1 - 2x_2 + 2x_3 = -1.$$

[*Ans.* $x_1 = 1, x_2 = 2, x_3 = 0$.]

(c)

$$3x_1 + 4x_2 + 5x_3 + x_4 = 8,$$

$$7x_1 + 2x_2 + 3x_3 + 2x_4 = 10,$$

$$4x_1 + 3x_2 + 2x_3 + x_4 = 6,$$

$$x_1 + 2x_2 + 3x_3 + x_4 = 4.$$

[*Ans.* $x_1 = 1, x_2 = 0, x_3 = 1, x_4 = 1$.]

2. Put each of the sets of equations (a), (b), and (c) in a canonical form. Determine their basic solutions, choosing as basic variables the ones indicated in each answer.

(a)

$$x_1 + x_2 + x_3 + x_4 = 100,$$

$$3x_1 + 2x_2 + 4x_3 + x_5 = 210,$$

$$3x_1 + 2x_2 + x_6 = 150.$$

[*Ans.* $x_4 = 47\frac{1}{2}$, $x_3 = 52\frac{1}{2}$, $x_5 = 150$.]

(b)

$$x + 4y - u = 24,$$

$$5x + y - v + b = 25,$$

$$x + 3y + pa + pb + M = 0.$$

[*Ans.* $a = 24$, $b = 25$, $M = -49p$.]

(c) Same as (b), but with x, a, and M as the basic variables.

[*Ans.* $a = 19$, $x = 5$, $M = -5 - 19p$.]

3. Test the following sets of equations for consistency and redundancy:

(a)

$$2x + 4y + 3z + w = 15,$$

$$3x + 7y + 2w = 16,$$

$$5x + 3y + 2z + 3w = 21.$$

[*Ans.* Consistent.]

(b)

$$4x + 8y + 6z + w = 18,$$

$$2x + 3y + 4z = 8,$$

$$7x + \tfrac{27}{2}y + 11z + \tfrac{3}{2}w = 31.$$

[*Ans.* Inconsistent.]

(c)

$$x_1 + x_2 + x_3 + x_4 = 10,$$

$$2x_1 + x_2 + 2x_3 = 10,$$

$$x_1 + 2x_2 + x_3 + 3x_4 = 20.$$

[*Ans.* Redundant.]

4. Why does the following system have more than one solution?

$$2x_1 + 3x_2 - 4x_3 = 8,$$

$$3x_1 - 2x_2 + 3x_3 = 4,$$

$$-6x_1 + 4x_2 - 6x_3 = -8.$$

[*Ans.* One equation is redundant; this means that there are two equations involving three unknowns.]

5. Test the following sets of equations for consistency:

(a)

$$2x_1 + x_2 + 4x_3 = 4,$$

$$x_1 - 3x_2 - x_3 = 5,$$

$$-3x_1 + 2x_2 - 2x_3 = 1,$$

$$8x_1 - 3x_2 + 8x_3 = 2.$$

[*Ans.* Consistent; $x_1 = 1$, $x_2 = 2$, $x_3 = 0$.]

(b)

$$2x_1 - x_2 + 4x_3 = 2,$$

$$x_1 + 2x_2 - 3x_3 = -4,$$

$$4x_1 + 3x_2 - 2x_3 = 6.$$

[*Ans.* Consistent, but last equation is redundant; $x_1 = -x_3$, $x_2 = 2x_3 - 2$.]

6. Show that the following set of equations is inconsistent:

$$x_1 + 2x_2 + 3x_3 + x_4 = 1,$$

$$x_1 - x_2 + 2x_3 - x_4 = -3,$$

$$3x_1 + 3x_2 + 8x_3 + x_4 = -3.$$

4

Simplex Method

4.1 SPECIFIC APPLICATION

In this chapter, we shall see how the Gauss-Jordan reduction process can be utilized to solve linear programming problems by the Simplex method. We shall introduce this method through the solution of two specific problems before generalizing the procedure. The first one that we shall consider is Problem 2.1 which dealt with the blending of teas. We repeat it here.

Problem 4.1 A tea company sells four brands of blended tea which we shall call brands 1, 2, 3, and 4. Each brand is a mixture, in varying proportions, of three types of tea leaves, A, B, and C. The amount of each of the three types in each kilogram of the four brands and the selling price per kilogram of each brand are shown in Table 4.1. A tea packing and export firm has discovered that each kilogram of the best blend should contain 0.42 kg of A, 0.27 kg of B, and 0.31 kg of C. How many kilograms of each of the four brands should the export firm buy in order to produce 1 kg of its own new brand while keeping its total purchasing cost at a minimum? How much is this cost?

TABLE 4.1

Type	*Brand*				*New Brand*
	1	2	3	4	
A	0.2	0.4	0.3	0.6	0.42
B	0.3	0.3	0.4	0.2	0.27
C	0.5	0.3	0.3	0.2	0.31
Selling Price per kg	\$ 2.25	\$ 3.00	\$ 2.50	\$ 3.25	Min Z

Let x_1, x_2, x_3, and x_4 be respectively the amounts of brands 1, 2, 3, and 4 in each kilogram of the new brand. The linear programming problem is:

$$\text{Minimize } Z = 2.25x_1 + 3x_2 + 2.5x_3 + 3.25x_4 \tag{4.1}$$

subject to

$$\begin{aligned} 0.2x_1 + 0.4x_2 + 0.3x_3 + 0.6x_4 &= 0.42, \\ 0.3x_1 + 0.3x_2 + 0.4x_3 + 0.2x_4 &= 0.27, \\ 0.5x_1 + 0.3x_2 + 0.3x_3 + 0.2x_4 &= 0.31, \\ x_1 + x_2 + x_3 + x_4 &= 1, \end{aligned} \tag{4.2}$$

$$x_1, x_2, x_3, x_4, \text{ all} \geqslant 0. \tag{4.3}$$

As noted in Problem 2.1, the last equation in (4.2) is redundant.

Now, any solution to Eqs. (4.2) that satisfies condition (4.3) is called a *feasible solution*; any solution to Eqs. (4.2) that satisfies Eq. (4.1) is called an *optimal solution*.

The problem is to determine a feasible solution that is at the same time optimal. A fundamental existence theorem in linear algebra states that *whenever there exists an optimal-feasible solution, it coincides with one of the basic-feasible solutions*. This is proved in Section 4.7. The Simplex method first determines the basic solutions to a given problem, then tests them for optimality. To determine these basic solutions, we adopt the Gauss-Jordan reduction process. To obtain the matrix of Eqs. (4.2) which will include Z too, we write the objective function as

$$-2.25x_1 - 3x_2 - 2.5x_3 - 3.25x_4 + Z = 0. \tag{4.4}$$

We write Eqs. (4.2) in a matrix form with detached coefficients, as in Tableau I. Choosing x_1, x_2, and x_4 as the basic variables, we shall reduce

TABLEAU I

x_1	x_2	x_3	x_4	Z	
0.2	0.4	0.3	0.6	0	0.42
0.3	0.3	0.4	0.2	0	0.27
0.5	0.3	0.3	0.2	0	0.31
−2.25	−3	−2.5	−3.25	1	0

this matrix to a canonical form. The succession of steps involved is shown in Tableau II, where the entries are made according to the steps detailed in Section 3.3. Further, Eq. (4.4) which includes Z is treated likewise, so that in the final table this equation does not contain the chosen basic variables x_1, x_2, and x_4. In the final step in Tableau II, the coefficients (including the last row representing the objective function) are in a

TABLEAU II

x_1	x_2	x_3	x_4	Z	
1	2	3/2	3	0	2.1
3	3	4	2	0	2.7
5	3	3	2	0	3.1
−9/4	−3	−5/2	−13/4	1	0
1	2	3/2	3	0	21/10
0	−3	−1/2	−7	0	−3.6
0	−7	−9/2	−13	0	−7.4
0	3/2	7/8	7/2	1	189/40
1	2	3/2	3	0	21/10
0	1	1/6	7/3	0	6/5
0	−7	−9/2	−13	0	−37/5
0	3/2	7/8	7/2	11	89/40
1	0	7/6	−5/3	0	−3/10
0	1	1/6	7/3	0	6/5
0	0	−10/3	10/3	0	1
0	0	5/8	0	1	117/40
1	0	7/6	−5/3	0	−3/10
0	1	1/6	7/3	0	6/5
0	0	−1	1	0	3/10
0	0	5/8	0	1	117/40
1	0	−1/2	0	0	1/5
0	1	5/2	0	0	1/2
0	0	−1	1	0	3/10
0	0	5/8 ↑	0	1	117/40

canonical form with respect to x_1, x_2, and x_4. From this, we can read the solutions as

$$
\begin{aligned}
x_1 &= 0.2 + 0.5x_3, \\
x_2 &= 0.5 - 2.5x_3, \\
x_4 &= 0.3 + x_3, \\
Z &= \frac{117}{40} - \frac{5}{8}x_3 = 2.925 - 0.625x_3.
\end{aligned}
\tag{4.5}
$$

If we put $x_3 \equiv 0$, the set of solutions obtained for x_1, x_2, and x_4 is a *basic solution.* This set satisfies the given set of equations in (4.2). Further, if these solutions for x_1, x_2, and x_4 satisfy conditions (4.3), i.e., the non-negative conditions, the set of solutions is called the *basic-feasible solution.* As Eqs. (4.5) show, the basic-feasible solution is

$$x_1 = 0.2, \qquad x_2 = 0.5, \qquad x_3 \equiv 0, \qquad x_4 = 0.3.$$

If we had chosen any other variables, say, x_2, x_3, and x_4, as our basic variables, we would have obtained a different basic solution with $x_1 \equiv 0$. However, this new basic solution would not be a basic-*feasible* solution unless these variables also satisfy conditions (4.3). Indeed, it will be shown in Section 4.2 that when we choose x_2, x_3, and x_4 as our basic variables, we will *not* get a *feasible* solution.

Coming back to the present problem, with x_1, x_2, and x_4 as basic variables and $x_3 \equiv 0$, we note that the value of the objective function Z is 2.925. However, Eqs. (4.5) show that since $Z = 2.925 - 0.625x_3$, the value of Z can be reduced if x_3 is given a non-zero (but positive) value. The pertinent question is: how much can the value of x_3 be increased? The answer is given by the first three equations in (4.5). From these equations, we observe:

(i) As x_3 is made non-zero, the values of x_1, x_2, and x_4 change.

(ii) x_1, x_2, and x_4 can be modified as long as conditions (4.3) are not violated, i.e., x_1, x_2, and x_4 remain non-negative.

(iii) The first and third equations in (4.5) show that x_3 can be increased infinitely without affecting the non-negativity requirements of x_1 and x_4. The second equation, however, puts a limit on x_3: here x_3 can be increased only up to 0.2; further increase would make x_2 negative.

(iv) The maximum value that x_3 can have is 0.2, and x_2 as a consequence becomes zero.

Statement (iv) indicates that x_3, instead of x_2, should become a basic variable so as to reduce the value of Z. The coefficient of x_3 that decided this change is shown in ***bold italic*** and is called the *pivot.* This means that the variable x_3 associated with this coefficient becomes the new basic variable replacing x_2, the earlier basic variable in that row. This replacement is again obtained by the Gauss-Jordan reduction, as shown in Tableau III. Such a tableau is generally called a *Simplex tableau.* In Tableau III, x_1, x_3, and x_4 appear as basic variables.

TABLEAU III

Basis	x_1	x_2	x_3	x_4	Z	
x_1	1	1/5	0	0	0	3/10
x_3	0	2/5	1	0	0	1/5
x_4	0	2/5	0	1	0	1/2
	0	−1/4	0	0	1	14/5

With the canonical form as in Tableau III, the solutions are

$$\begin{aligned} x_1 &= 0.3 - 0.2x_2, \\ x_3 &= 0.2 - 0.4x_2, \\ x_4 &= 0.5 - 0.4x_2, \\ Z &= 2.8 + 0.25x_2. \end{aligned} \tag{4.6}$$

The objective function Z will have a minimum value when $x_2 = 0$, since the coefficient of x_2 is positive. This minimum value, which is 2.8, is also the absolute minimum for Problem 4.1. This conclusion is based on the fundamental theorem (which will be proved in Section 4.6) that when all the coefficients of the non-basic variables in the objective function (expressed in the canonical form) are positive, the value obtained for Z, when all these non-basic variables are equated to zero, is the optimal value. In Eqs. (4.6), the coefficient of the non-basic variable x_2 in the objective function Z is positive. If a non-zero value is given to x_2, the value of Z will increase. Hence, putting $x_2 \equiv 0$, the objective function becomes 2.8 which is the minimum. Hence, in the tea-blending problem, $x_1 = 0.3$ kg, $x_3 = 0.2$ kg, and $x_4 = 0.5$ kg. These are respectively the amounts of brands 1, 3, and 4 that the export firm should buy. Brand 2 is not purchased. The purchasing cost per kilogram of the best blend is \$ 2.8. If the firm had purchased 0.3 kg of brand 4, as Tableau II shows, the mixture would still be satisfactory, but its cost would have been \$ 2.925 per kilogram, as shown in Eqs. (4.5).

4.2 OTHER BASIC SOLUTIONS

In the foregoing analysis, we chose x_1, x_2, and x_4 as our first basic variables. These gave solutions that were also feasible, i.e., $x_1 \geqslant 0$, $x_2 \geqslant 0$, and $x_4 \geqslant 0$. This initial set led to a second set of basic variables x_1, x_3, and x_4 which also were feasible. If we had chosen x_2, x_3, and x_4 or x_1, x_2, and x_3 as the set of basic variables, Eqs. (4.2) would have yielded the solutions

$$(x_2, x_3, x_4) \equiv (1.5, -0.4, -0.1),$$

$$(x_1, x_2, x_3) \equiv (0.05, 1.25, -0.3).$$

As these values do not satisfy the non-negativity requirements, they are not basic-feasible solutions. This raises the important question of finding a method to determine an *initial* basic-feasible solution. The Simplex tableau will then lead us to other basic-feasible solutions and the optimal solution (assuming it exists). The method of finding the *initial basic-feasible solution* will be discussed in Chapter 5.

We shall demonstrate, through the solution of Problem 4.2, some of the important points discussed so far.

Problem 4.2

$$\text{Minimize } Z = 4x_1 + 2x_2 + 3x_3$$

subject to the constraints

$$2x_1 + 4x_3 \geqslant 5,$$

$$2x_1 + 3x_2 + x_3 \geqslant 4,$$

$$x_1, x_2, x_3, \text{ all} \geqslant 0.$$

By adding slack variables, we can put this problem in the standard form:

$$\text{Minimize } Z = 4x_1 + 2x_2 + 3x_3 \tag{4.7}$$

subject to

$$\begin{aligned} 2x_1 + 4x_3 - x_4 &= 5, \\ 2x_1 + 3x_2 + x_3 - x_5 &= 4, \end{aligned} \tag{4.8}$$

$$x_1, x_2, x_3, x_4, x_5, \text{ all} \geqslant 0. \tag{4.9}$$

We shall solve this problem in a manner slightly different from that in Problem 4.1 in order to make certain points clear before applying the Simplex tableau.

There are only two equations in (4.8) and the basis will consist of two variables from among the five admissible variables. There can be ten possible combinations, for instance, (x_1, x_2), and (x_1, x_3). Let us, for example, choose (x_1, x_3) as the basic variables, and solve the values of Eqs. (4.8) while equating the remaining non-basic variables x_2, x_4, and x_5 to zero. We get

$$x_1 = 11/6, \qquad x_3 = 1/3, \qquad Z = 25/3.$$

Similarly, choosing other bases, we can determine their values and the corresponding values of the objective function as in Table 4.2. As the values show, some of these bases do not satisfy the non-negativity requirement imposed by conditions (4.9). Hence, these basic variables are not feasible. The basic-feasible variables (marked with asterisks) are (x_1, x_3), (x_1, x_5), (x_2, x_3), and (x_3, x_4). The minimum Z occurs for (x_2, x_3), with $Z_{\min} = 67/12$.

TABLE 4.2

Basis	*Value*	*Z*	*Basis*	*Value*	*Z*
(x_1, x_2)	$(5/2, -1/3)$	28/3	(x_2, x_4)	$(4/3, -5)$	16/3
$*(x_1, x_3)$	$(11/6, 1/3)$	25/3	(x_2, x_5)	No solution	
(x_1, x_4)	$(2, -1)$	8	$*(x_3, x_4)$	$(4, 11)$	12
$*(x_1, x_5)$	$(5/2, 11)$	10	(x_3, x_5)	$(5/4, -11/4)$	5
$*(x_2, x_3)$	$(11/12, 5/4)$	67/12	(x_4, x_5)	$(-5, -4)$	0

The *Simplex method* produces the same result. If we choose (x_2, x_3) as the basis, the solution is obtained in one step. But in order to get some practice, we shall select (x_1, x_3) as the initial basis in Tableau I.

TABLEAU I

Basis	x_1	x_2	x_3	x_4	x_5	Z	
	2	0	4	-1	0	0	5
	2	3	1	0	-1	0	4
	-4	-2	-3	0	0	1	0
	1	0	2	$-1/2$	0	0	5/2
	0	3	-3	1	-1	0	-1
	0	-2	5	-2	0	1	10
x_1	1	2	0	1/6	$-2/3$	0	11/6
x_3	0	-1	1	$-1/3$	1/3	0	1/3
	0	3 ↑	0	$-1/3$	$-5/3$	1	25/3

With the given Eqs. (4.8) reduced to a canonical form, as in Tableau I, the solutions are

$$\begin{aligned} x_1 &= 11/6 - 2x_2 - (1/6)x_4 + (2/3)x_5, \\ x_3 &= 1/3 + x_2 + (1/3)x_4 - (1/3)x_5, \\ Z &= 25/3 - 3x_2 + (1/3)x_4 + (5/3)x_5. \end{aligned} \tag{4.10}$$

With the non-basic variables x_2, x_4, and x_5 equated to zero, the initial basic-feasible solution is

$$x_1 = 11/6, \qquad x_3 = 1/3, \qquad Z = 25/3.$$

However, as Eqs. (4.10) show, the value of Z can be made less than 25/3 if x_2 is made non-zero. In order to find the maximum value that x_2 can assume, let us examine the first two of Eqs. (4.10). With $x_4 = x_5 \equiv 0$, x_2 can have a maximum value of 11/12 without making x_1 non-negative. The non-negativity of x_3 is unaffected by the value of x_2. When x_2 is made equal to 11/12, x_1 becomes zero, i.e., x_2 is the new basic variable, replacing x_1. The coefficient of x_2 which is 2 is accordingly shown in ***bold italic*** to indicate that it is the pivot. Taking (x_2, x_3) as the basis, we get Tableau II.

TABLEAU II

Basis	x_1	x_2	x_3	x_4	x_5	Z	
x_2	1/2	1	0	1/12	−1/3	0	11/12
x_3	1/2	0	1	−1/4	0	0	5/4
	−3/2	0	0	−7/12	−2/3	1	67/12

The coefficients in the last row are either negative or zero; hence, the value of Z cannot be reduced any further. This is because

$$Z = 67/12 + (3/2)x_1 + (7/12)x_4 + (2/3)x_5$$

and Z has the minimum value when $x_1 = 0$, $x_4 = 0$, and $x_5 = 0$. The optimal solutions are therefore

$$x_2 = 11/12, \qquad x_3 = 5/4, \qquad x_1 = x_4 = x_5 \equiv 0,$$

$$Z_{\min} = 67/12.$$

4.3 GENERAL PROCEDURE

Let the linear programming problem be:

$$\text{Minimize } Z = c_1x_1 + c_2x_2 + \ldots + c_nx_n \tag{4.11}$$

subject to

$$\begin{aligned} a_{11}x_1 + a_{12}x_2 + \ldots + a_{1n}x_n &= b_1 \\ a_{21}x_1 + a_{22}x_2 + \ldots + a_{2n}x_n &= b_2 \\ &\vdots \\ a_{m1}x_1 + a_{m2}x_2 + \ldots + a_{mn}x_n &= b_m \end{aligned} \tag{4.12}$$

$$x_1 \geqslant 0, x_2 \geqslant 0, \ldots, x_n \geqslant 0. \tag{4.13}$$

Let us arbitrarily select $x_1, x_2, \ldots, x_m$ as the basic variables and express the constraint equations in a canonical form with respect to these variables, as follows:

$$\begin{aligned} x_1 \qquad\qquad + \bar{a}_{1,m+1}x_{m+1} + \ldots + \bar{a}_{1n}x_n &= \bar{b}_1 \\ x_2 \qquad + \bar{a}_{2,m+1}x_{m+1} + \ldots + \bar{a}_{2n}x_n &= \bar{b}_2 \\ &\vdots \\ x_m + \bar{a}_{m,m+1}x_{m+1} + \ldots + \bar{a}_{mn}x_n &= \bar{b}_m \end{aligned} \tag{4.14}$$

From Eqs. (4.14), we get

$$\begin{aligned} x_1 &= \bar{b}_1 - \bar{a}_{1,m+1}x_{m+1} - \ldots - \bar{a}_{1n}x_n = \bar{b}_1 - \sum_{r=m+1}^{n} \bar{a}_{1r}x_r \\ x_2 &= \bar{b}_2 - \bar{a}_{2,m+1}x_{m+1} - \ldots - \bar{a}_{2n}x_n = \bar{b}_2 - \sum_{r=m+1}^{n} \bar{a}_{2r}x_r \\ &\vdots \\ x_m &= \bar{b}_m - \bar{a}_{m,m+1}x_{m+1} - \ldots - \bar{a}_{mn}x_n = \bar{b}_m - \sum_{r=m+1}^{n} \bar{a}_{mr}x_r \end{aligned} \tag{4.15}$$

Substituting these in the objective function Z, we get

$$Z = c_1\bar{b}_1 + c_2\bar{b}_2 + \ldots + c_m\bar{b}_m - c_1 \sum_{r=m+1}^{n} \bar{a}_{1r}x_r - c_2 \sum_{r=m+1}^{n} \bar{a}_{2r}x_r$$

$$- \ldots - c_m \sum_{r=m+1}^{n} \bar{a}_{mr}x_r + c_{m+1}x_{m+1} + \ldots + c_nx_n$$

or

$$Z = \bar{Z} + \bar{c}_{m+1}x_{m+1} + \ldots + \bar{c}_nx_n, \tag{4.16}$$

where

$$\bar{c}_{m+1} = c_{m+1} - \sum_{r=1}^{m} c_r\bar{a}_{r,m+1}$$

$$\bar{c}_{m+2} = c_{m+2} - \sum_{r=1}^{m} c_r\bar{a}_{r,m+2}$$

$$\vdots$$

$$\bar{c}_n = c_n - \sum_{r=1}^{m} c_r\bar{a}_{r,n}$$

$$\bar{Z} = \sum_{r=1}^{m} c_r\bar{b}_r.$$

When the non-basic variables are set equal to zero, the basic solution that immediately results from Eqs. (4.15) and (4.16) is

$$x_1 = \bar{b}_1,\ x_2 = \bar{b}_2,\ \ldots,\ x_m = \bar{b}_m,$$

$$x_{m+1} = x_{m+2} = \ldots = x_n \equiv 0,$$

$$Z = \bar{Z}.$$

If $\bar{b}_1, \bar{b}_2, \ldots, \bar{b}_m$ are non-negative, the basic solution is also a feasible solution. If one or more $\bar{b}_r$'s $(r = 1, 2, \ldots, m)$ are zero, the basic solution is said to be *degenerate*, although the b_r's are still non-negative. For the present, let us assume that we have a basic-feasible solution that is not degenerate. The value of the objective function corresponding to the present basic-feasible solution is $\bar{Z}$. The question is how to decide whether or not $Z = \bar{Z}$ thus obtained is a minimum.

The answer to this question lies in examining how Z in Eq. (4.16) varies when we move away from the original basis. If all the $\bar{c}_j$'s $(j = m + 1, m + 2, \ldots, n)$ are non-negative, we cannot make the value of Z less than that of $\bar{Z}$, since all the x_j's $(j = m + 1, m + 2, \ldots, n)$ are non-negative. On the other hand, if one or more of the coefficients, say, $\bar{c}_p, \bar{c}_q, \ldots$, are negative, the value of Z can be made less than that of $\bar{Z}$ by making one or all of the variables $x_p, x_q, \ldots$ associated with these coefficients non-zero. However, according to the fundamental theorem (which is proved in Section 4.6), the optimal solution (if it exists) corresponds to one of the basic solutions. Hence, in the process of reducing Z further from $\bar{Z}$, we should not make all the variables $x_p, x_q, \ldots$ with negative coefficients non-zero, but choose one of them as the basic variable that replaces an existing one. The natural choice for the new basic variable will be the non-basic variable that has the largest negative coefficient, say, x_p, since, when it forms the new basis, Z can be reduced fastest. In order to determine which of the present basic variables should become non-basic, we observe the following from Eqs. (4.15).

When one of the non-basic variables, x_p, is made non-zero, the solutions for the basic variables change from the present values $\bar{b}_1, \bar{b}_2, \ldots$, as shown in Eqs. (4.15). In this process, we should take care to see that none of the existing basic variables becomes non-negative. If all the coefficients of x_p in Eqs. (4.15), i.e., $\bar{a}_{1p}, \bar{a}_{2p}, \ldots$, are positive, then x_p can be increased to any limit without making any of the basic variables non-negative. In such a case, the solution is unbounded. On the other hand, if one or more of the coefficients of x_p in Eqs. (4.15) are negative, we choose the one that gives a minimum value for x_p while making one of the present variables zero without making it negative. This is the basic variable that will become non-basic, yielding its place to x_p. The coefficient of x_p that decided this replacement is called the *pivot*.

With the new basic variable x_p in place of one of the original ones, and the others remaining unchanged, we reduce the equations to a canonical form to determine the new objective function. This is the second iteration and the table representing it is called *Simplex Tableau II*. The objective function is tested for its optimality. If all the coefficients are positive, the optimal value is the corresponding $\bar{Z}$. Otherwise, the iteration is continued until the optimal value is obtained.

The optimality test that we applied in this minimization problem to see whether or not the value of Z obtained in one of the iteration processes is an optimal value is generally called the *Simplex criterion*. According to this criterion, if the coefficients of all non-basic values as they appear in the Simplex tableau are negative, the optimal solution is reached and $Z_{\min} = \bar{Z}$.

If the optimization problem deals with the maximization of the objective function, we need not convert it into a minimization problem. We can leave it as it is and apply the Simplex procedure. However, in this case, the Simplex criterion will determine that the optimal (i.e., maximum) solution is reached when the coefficients of all the non-basic variables in the Simplex tableau are *positive*. If some of the coefficients are negative, we would need to choose the *most negative* of these and continue the iteration.

4.4 BASIC STEPS

We shall recapitulate the foregoing steps and indicate how the entries are made in a tabular form.

(i) Let us assume that we have arbitrarily chosen $x_1, x_2, \ldots, x_m$ as the basic variables yielding a feasible solution. Simplex Tableau I will also contain the objective function Z in the form

$$-\bar{c}_{m+1}x_{m+1} - \bar{c}_{m+2}x_{m+2} - \ldots - \bar{c}_n x_n + Z = \bar{Z}.$$

SIMPLEX TABLEAU I

Basis	x_1	$\ldots$	x_k	$\ldots$	x_m	x_{m+1}	$\ldots$	x_r	$\ldots$	x_n	Z	
x_1	1					$\bar{a}_{1,m+1}$	$\ldots$	$\bar{a}_{1r}$	$\ldots$	$\bar{a}_{1n}$		$\bar{b}_1$
$\vdots$												
x_k			1			$\bar{a}_{k,m+1}$	$\ldots$	$\bar{a}_{kr}$	$\ldots$	$\bar{a}_{kn}$		$\bar{b}_k$
$\vdots$												
x_m					1	$\bar{a}_{m,m+1}$	$\ldots$	$\bar{a}_{mr}$	$\ldots$	$\bar{a}_{mn}$		$\bar{b}_m$
						$-\bar{c}_{m+1}$	$\ldots$	$-\bar{c}_r$ $\uparrow$	$\ldots$	$-\bar{c}_n$	1	$\bar{Z}$

(ii) If all the entries in the last row, i.e., $(-\bar{c}_{m+1}), (-\bar{c}_{m+2}), \ldots, (-\bar{c}_n)$, are *negative*, the value of $Z = \bar{Z}$ is the optimal value. Otherwise, we choose the *most positive* among these entries. Let this be $(-\bar{c}_r)$, indicated by an arrow. The variable associated with this coefficient, i.e., x_r, becomes the new basic variable. (If the problem deals with maximizing, we can leave it as it is without converting it into a minimization problem. In a case that requires maximization, we choose the *most negative* coefficient in order to apply the Simplex criterion.)

(iii) If all the coefficients $\bar{a}_{1r}, \bar{a}_{2r}, \ldots, \bar{a}_{mr}$ in the column x_r are negative, the solution is unbounded. On the other hand, if some are positive, we select the one among $\bar{b}_{1/\bar{a}_{1r}}, \bar{b}_{2/\bar{a}_{2r}}, \ldots, b_{m/\bar{a}_{mr}}$ that gives the minimum value of x_r and show it in ***bold italic*** to denote it is the pivot. Let this be $\bar{a}_{kr}$. The basic variable x_k, in the second row, now becomes non-basic, yielding its place to x_r.

(iv) We write Simplex Tableau II, with x_r as the new basic variable in place of x_k.

SIMPLEX TABLEAU II

Basis	x_1	...	x_k	...	x_m	x_{m+1}	...	x_r	...	x_n	Z	
x_1	1		$\bar{a}'_{1,k}$			$\bar{a}'_{1,m+1}$	...	0	...	$\bar{a}'_{1n}$		$\bar{b}'_1$
⋮												
x_r			$\bar{a}'_{k,k}$			$\bar{a}'_{k,m+1}$	...	1	...	$\bar{a}'_{kn}$		$\bar{b}'_k$
⋮												
x_m			$\bar{a}'_{m,k}$		1	$\bar{a}'_{m,m+1}$	...	0	...	$\bar{a}'_{mn}$		$\bar{b}'_m$
			$-\bar{c}'_k$			$-\bar{c}'_{m+1}$	...	0	...	$-\bar{c}'_n$	1	$\bar{Z}'$

(v) Applying the Simplex criterion to the last row, we determine whether $\bar{Z}'$ is the optimal value of Z. If not, we repeat Steps (ii) to (iv).

Let us apply Steps (i) to (v) to Problem 2.2 which dealt with the dairy-milk distributor.

Problem 4.3 The standard form of Problem 2.2 is:

$$\text{Minimize } Z^* = -Z = -3x_1 - 4x_2 - 5x_3$$

subject to

$$x_1 + x_2 + x_3 + x_4 = 120,$$

$$x_1 + 2x_2 + 3x_3 + x_5 = 150,$$

$$x_1 + 2x_2 + x_6 = 90,$$

$$x_1 \geqslant 0, x_2 \geqslant 0, \ldots, x_6 \geqslant 0.$$

The equations are already in a canonical form with respect to the variables x_4, x_5, and x_6. Thus, Tableau I can be written as shown.

TABLEAU I

Basis	x_1	x_2	x_3	x_4	x_5	x_6	Z	
x_4	1	1	1	1				120
x_5	1	2	*3*		1			150
x_6	1	2				1		90
	3	4	5				−1	0
			↑					

The most positive of the entries in the last row in Tableau I is 5. We then select the minimum between 120/1 and 150/3. The coefficient 3 corresponding to the minimum value is the pivot. x_3 is therefore the new basic variable (replacing x_5) in Tableau II.

TABLEAU II

Basis	x_1	x_2	x_3	x_4	x_5	x_6	Z	
x_4	2/3	1/3		1	−1/3			70
x_3	1/3	2/3	1		1/3			50
x_6	*1*	2				1		90
	4/3	2/3			−5/3		−1	−250
	↑							

As Tableau III indicates, the new basic variable x_1 replaces the variable x_6.

TABLEAU III

Basis	x_1	x_2	x_3	x_4	x_5	x_6	Z	
x_4		−1		1	−1/3	−2/3		10
x_3			1		1/3	−1/3		20
x_1	1	2				1		90
		−2			−5/3	−4/3	−1	−370

Now, all the entries in the last row in Tableau III are negative; hence, the optimal value for $-Z = -370$. Therefore, in the milk distribution problem, $x_1 = 90$, $x_2 = 0$, and $x_3 = 20$. These are respectively the number of bottles the distributor supplies to areas A, B, and C. He earns \$ 3.70

per day which is the optimal earning. Though he can carry 120 bottles, he can supply only 110 bottles—the slack variable x_4 is therefore equal to 10—, because he can spare only 150 minutes for this milk distribution job, and this time has been spent during the delivery of 90 bottles to area A (=90 minutes) and 20 bottles to area C (=60 minutes).

Instead of converting the problem into a minimization problem, if we had left it as

$$\text{maximize } Z = 3x_1 + 4x_2 + 5x_3,$$

the only difference would have been that, in the last row of each tableau, the coefficients would have had signs the opposite of the present ones. The Simplex criterion in such a case would have suggested that we select the most negative coefficient to determine the new basic variable. The rest of the procedure would remain unaltered.

4.5 DEGENERACY

In "well-behaved" linear programming problems, all the basic solutions as given by the $\bar{b}_r$'s during the several iterative steps will be positive. There may occur, however, a few cases in which one or more of the basic variables become zero at some stage of the calculations. In such cases, the basic solution is said to be *degenerate*. We shall now discuss the two main effects of a degenerate case:

(i) During the iterative process, one of the two available variables has to be selected to yield place to a new basic variable.

(ii) There is the slight possibility of the iteration resulting in a situation called *cycling*.

Degeneracy may occur at the first step, when some of the constraint conditions are in the form of equalities, with the slack variables in these equations being equal to zero. Or, degeneracy may occur at a later stage during iteration, when two or more basic variables in the table set the same effective limit to the new basic variable. To make this clear, let us consider Problem 4.4.

Problem 4.4

$$\text{Maximize } Z = 7x_1 + 12x_2 + 16x_3$$

subject to

$$2x_1 + x_2 + x_3 + x_4 = 1,$$

$$x_1 + 2x_2 + 4x_3 + x_5 = 2,$$

$$x_1, x_2, x_3, x_4, x_5, \text{ all} \geqslant 0.$$

Based on the given data, we formulate Tableaux I and II.

TABLEAU I

Basis	x_1	x_2	x_3	x_4	x_5	Z	
x_4	2	1	1	1			1
x_5	1	2	*4*		1		2
	−7	−12	−16			1	0
			↑				

TABLEAU II

Basis	x_1	x_2	x_3	x_4	x_5	Z	
x_4	7/4	1/2		1	−1/4		1/2
x_3	1/4	*1/2*	1		1/4		1/2
	−3	−4			4	1	8
		↑					

When we examine Tableau II, we find that both the existing basic variables x_3 and x_4 set the same limit to the new variable x_2. Thus, both x_3 and x_4 become zero when x_2 is increased to the maximum value compatible with the non-negativity requirement. However, in order to continue the iteration, since the optimal solution has not yet been obtained, we arbitrarily select x_3 as the replaced variable and continue the iteration, as shown in Tableau III.

TABLEAU III

Basis	x_1	x_2	x_3	x_4	x_5	Z	
x_4	*3/2*		−1	1	−1/2		0
x_2	1/2	1	2		1/2		1
	−1		8		6		12
	↑						

The last row in Tableau III shows that we have not yet reached the optimal solution. The replaced variable is x_4 (because the minimum between 0/(3/2) and 1/(1/2) corresponds to this). We proceed formally to Tableau IV.

We observe that both Tableaux III and IV are degenerate, and in each case Z is equal to 12. Though this value of Z is the optimal value, the entry in Tableau III does not satisfy the Simplex criterion. We therefore proceeded to Tableau IV, where the iteration satisfies the Simplex criterion but results in the same solution as in Tableau III, that is, $Z_{max} = 12$ and

TABLEAU IV

Basis	x_1	x_2	x_3	x_4	x_5	Z
x_1	1		$-2/3$	$2/3$	$-1/3$	0
x_2		1	$7/3$	$-1/3$	$2/3$	1
			$22/3$	$2/3$	$17/3$	12

$x_2 = 1$. This establishes the important fact that the Simplex criterion is not a *necessary*, though always a *sufficient*, condition for optimality.

The danger in a degenerate problem is in the slight possibility of its creating a situation called cycling. In such a case, after going through several iterative steps, we come back to a basis that was earlier rejected. This means that the procedure will repeat itself without going through an optimal stage. Such a non-terminating situation rarely occurs in practice. Indeed, it is difficult to construct even a hypothetical problem illustrating cycling. Nevertheless, if such a situation occurs, it can be overcome by a simple rule; however, we shall not go further into this aspect, because of its rare occurrence.

4.6 FUNDAMENTAL THEOREM

In the solutions to Problems 4.1 to 4.4, we have been explicitly assuming that the optimal solution coincided with one of the basic-feasible solutions. We shall now prove this assumption.

Consider the linear programming problem in n admissible variables (including the slack variables):

Determine $x_1, x_2, \ldots, x_n$

subject to

$$
\begin{aligned}
a_{11}x_1 + a_{12}x_2 + \ldots + a_{1n}x_n &= b_1 \\
a_{21}x_1 + a_{22}x_2 + \ldots + a_{2n}x_n &= b_2 \\
&\vdots \\
a_{m1}x_1 + a_{m2}x_2 + \ldots + a_{mn}x_n &= b_m \\
x_1 \geqslant 0,\ x_2 \geqslant 0, \ldots, x_n &\geqslant 0
\end{aligned}
\tag{4.17}
$$

so as to

$$\text{minimize } Z = c_1x_1 + c_2x_2 + \ldots + c_nx_n.$$

We are interested in the case where $n > m$. According to the *fundamental theorem*, if a set of equations such as Eqs. (4.17) can be solved, there will be an optimal solution in which, *at most*, m of the admissible variables will be positive. This is to say that there will be at least one basic-feasible solution and the optimal solution will coincide with that particular basic-feasible solution or with one of the basic-feasible solutions (if more than one exist).

To prove this, let us assume that Eqs. (4.17) can be solved and that there are p number of admissible variables that are positive. Let the first p variables be $x_1, x_2, \ldots, x_p$. The rest of the $(n-p)$ variables are zero. If p happens to be less than m, the theorem is automatically satisfied. We have to show that p cannot exceed m. Let us assume that $p > m$ and examine its implication. Let the solution be

$$x_1 = \lambda_1, x_2 = \lambda_2, \ldots, x_p = \lambda_p, x_{p+1} = 0, \ldots, x_n = 0.$$

This satisfies

$$\begin{aligned} &a_{11}\lambda_1 + a_{12}\lambda_2 + \ldots + a_{1p}\lambda_p = b_1 \\ &a_{21}\lambda_1 + a_{22}\lambda_2 + \ldots + a_{2p}\lambda_p = b_2 \\ &\vdots \\ &a_{m1}\lambda_1 + a_{m2}\lambda_2 + \ldots + a_{mp}\lambda_p = b_m \\ &\lambda_1 > 0, \lambda_2 > 0, \ldots, \lambda_p > 0, \\ &c_1\lambda_1 + c_2\lambda_2 + \ldots + c_p\lambda_p = Z_{\min}. \end{aligned} \tag{4.18}$$

Now, let us consider the set of homogeneous (i.e., all the right-hand side members equal zero) equations

$$\begin{aligned} &a_{11}\beta_1 + a_{12}\beta_2 + \ldots + a_{1p}\beta_p = 0 \\ &a_{21}\beta_1 + a_{22}\beta_2 + \ldots + a_{2p}\beta_p = 0 \\ &\vdots \\ &a_{m1}\beta_1 + a_{m2}\beta_2 + \ldots + a_{mp}\beta_p = 0 \end{aligned} \tag{4.19}$$

In Eqs. (4.19), the number of variables $\beta_1, \beta_2, \ldots, \beta_p$ exceeds the number of equations. Using the Gauss-Jordan reduction process, we get

$$\begin{aligned} &\beta_1 + \bar{a}_{1,m+1}\beta_{m+1} + \ldots + \bar{a}_{1p}\beta_p = 0 \\ &\beta_2 + \bar{a}_{2,m+1}\beta_{m+1} + \ldots + \bar{a}_{2p}\beta_p = 0 \\ &\vdots \\ &\beta_m + \bar{a}_{m,m+1}\beta_{m+1} + \ldots + \bar{a}_{mp}\beta_p = 0 \end{aligned}$$

Solving for $\beta_1, \beta_2, \ldots,$ we have

$$\begin{aligned} &\beta_1 = -\bar{a}_{1,m+1}\beta_{m+1} - \ldots - \bar{a}_{1p}\beta_p \\ &\beta_2 = -\bar{a}_{2,m+1}\beta_{m+1} - \ldots - \bar{a}_{2p}\beta_p \\ &\vdots \\ &\beta_m = -\bar{a}_{m,m+1}\beta_{m+1} - \ldots - \bar{a}_{mp}\beta_p \end{aligned}$$

Depending on the values assigned to $\beta_{m+1}, \beta_{m+2}, \ldots, \beta_p$, the left-hand side members will have an infinite set of solutions. In other words, Eqs. (4.19) can be solved where at least one β has a non-zero value.

(There is obviously a trivial solution to the equations where $\beta_1 = \beta_2 = \ldots = \beta_p \equiv 0$. In addition, there exist other solutions in which at least one β is not equal to zero.) Among these non-zero β's, let there be at least one that is positive. Otherwise, we can multiply Eqs. (4.19) by (-1) to give at least one positive β and the equations would be satisfied. Further, among these β's, let β_r be the maximum, and among the λ's, λ_r the minimum so that $\theta = \beta_r/\lambda_r$ is the maximum among all the numbers

$$\beta_r/\lambda_1, \beta_r/\lambda_2, \ldots, \beta_r/\lambda_p.$$

Clearly, θ is a positive number. Dividing all the terms in Eqs. (4.19) by θ and subtracting these from Eqs. (4.18), we get

$$a_{11}(\lambda_1 - \frac{\beta_1}{\theta}) + a_{12}(\lambda_2 - \frac{\beta_2}{\theta}) + \ldots + a_{1p}(\lambda_p - \frac{\beta_p}{\theta}) = b_1$$

$$a_{21}(\lambda_1 - \frac{\beta_1}{\theta}) + a_{22}(\lambda_2 - \frac{\beta_2}{\theta}) + \ldots + a_{2p}(\lambda_p - \frac{\beta_p}{\theta}) = b_2$$

$$\vdots$$

$$a_{m1}(\lambda_1 - \frac{\beta_1}{\theta}) + a_{m2}(\lambda_2 - \frac{\beta_2}{\theta}) + \ldots + a_{mp}(\lambda_p - \frac{\beta_p}{\theta}) = b_m$$

This new set of equations shows that the set of numbers

$$(\lambda_1 - \frac{\beta_1}{\theta}), (\lambda_2 - \frac{\beta_2}{\theta}), \ldots, (\lambda_p - \frac{\beta_p}{\theta})$$

is also a solution to Eqs. (4.17) and is non-negative, because

$$\lambda_j > 0, \qquad \theta > 0, \qquad \lambda_j \geqslant \lambda_r, \qquad \beta_j \leqslant \beta_r,$$

$$\lambda_j - \frac{\beta_j}{\theta} = \frac{\lambda_j\beta_r - \lambda_r\beta_j}{\beta_r} \geqslant 0 \qquad (j = 1, 2, \ldots, p).$$

Further, at least one of the numbers (when $j = r$) is zero. This means that we have found a feasible solution in which less than p variables are positive. We shall now show that this new set also gives an optimal value, i.e.,

$$c_1(\lambda_1 - \frac{\beta_1}{\theta}) + c_2(\lambda_2 - \frac{\beta_2}{\theta}) + \ldots + c_p(\lambda_p - \frac{\beta_p}{\theta})$$
$$= c_1\lambda_1 + c_2\lambda_2 + \ldots + c_p\lambda_p = Z_{\min}.$$

This is true when

$$c_1\lambda_1 + c_2\lambda_2 + \ldots + c_p\lambda_p = 0. \tag{4.20}$$

Let us assume that Eq. (4.20) is not true. Then we can find a number μ such that

$$\mu(c_1\beta_1 + c_2\beta_2 + \ldots + c_p\beta_p) < 0,$$

i.e.,

$$c_1(\mu\beta_1) + c_2(\mu\beta_2) + \ldots + c_p(\mu\beta_p) < 0.$$

Now, adding $c_1\lambda_1 + c_2\lambda_2 + \ldots + c_p\lambda_p$ to both sides, we get

$$c_1(\lambda_1 + \mu\beta_1) + c_2(\lambda_2 + \mu\beta_2) + \ldots + c_p(\lambda_p + \mu\beta_p)$$
$$< c_1\lambda_1 + c_2\lambda_2 + \ldots + c_p\lambda_p = Z_{\min}. \tag{4.21}$$

The set of numbers

$$(\lambda_1 + \mu\beta_1), (\lambda_2 + \mu\beta_2), \ldots, (\lambda_p + \mu\beta_p) \tag{4.22}$$

satisfies Eqs. (4.18), because the set $\mu\beta_1, \mu\beta_2, \ldots, \mu\beta_p$ satisfies the homogeneous set of Eqs. (4.19) for any value of $|\mu|$. By making $|\mu|$ sufficiently small, all the numbers in (4.22) can be made non-negative. This means that the set of numbers in (4.22) would give for Z a value less than $Z_{\min}$, according to Eq. (4.21). This clearly contradicts our assumption that $\lambda_1, \lambda_2, \ldots, \lambda_p$ give the optimal solution. Hence, Eq. (4.20) must be true.

We have therefore proved that the number of positive variables giving an optimal value can always be reduced so long as $p > m$. Repeating this argument eventually leads to an optimal solution where, at most, m admissible variables are positive.

4.7 SIMPLEX CRITERION

We have been applying the Simplex criterion to determine whether or not the objective function during the iterative procedure had reached its optimal value. The logic in its application was fairly elementary. Now, having proved the fundamental theorem, we can give a complete explanation for the criterion.

The fundamental theorem states that if an optimal solution exists for a linear programming problem, as stated by (4.17), it will coincide with one of the basic-feasible solutions. With m independent constraint equations involving n admissible variables, let us assume that we have found one basic-feasible solution. Let the basic variables be $x_1, x_2, \ldots, x_m$. Reducing the constraint equations to a canonical form, we get (from Section 4.4)

$$\begin{aligned} x_1 &= \bar{b}_1 - \bar{a}_{1,m+1}x_{m+1} - \ldots - \bar{a}_{1,n}x_n \\ x_2 &= \bar{b}_2 - \bar{a}_{2,m+1}x_{m+1} - \ldots - \bar{a}_{2,n}x_n \\ &\vdots \\ x_m &= \bar{b}_m - \bar{a}_{m,m+1}x_{m+1} - \ldots - \bar{a}_{m,n}x_n \end{aligned} \tag{4.23}$$

$$Z = \bar{Z} + c_{m+1}x_{m+1} + \ldots + c_nx_n. \tag{4.24}$$

The basic-feasible solution is

$$x_1 = \bar{b}_1,\ x_2 = \bar{b}_2,\ \ldots,\ x_m = \bar{b}_m,$$

$$Z = \bar{Z}.$$

With this basic-feasible solution, the value of the objective function is $\bar{Z}$. This is on the basis that $x_{m+1}, x_{m+2}, \ldots, x_n$ are equated to zero in Eq. (4.24). However, if by making one or more of these variables non-zero, the value of Z is reduced (in the case of a minimization problem) from its present value of $\bar{Z}$, then obviously $\bar{Z}$ is not an optimal value. Further, the non-basic variables in Eq. (4.24) that give this improvement in the value of $\bar{Z}$ are those with negative coefficients. But, according to the fundamental theorem, we have to find the optimal solution among the basic-feasible solutions. Hence, we change the value of only one non-basic variable from its present zero value and use this new variable in the place of an existing one.

To get the fastest improvement in the value of Z, we select the non-basic variable in Eq. (4.24) that has the most negative coefficient. Let this variable be x_r. In the process of improving Z, we cannot give an unlimited value to this non-basic variable, because, according to Eqs. (4.23), x_1, $x_2, \ldots, x_m$ have to remain non-negative. If all the coefficients of x_r in Eqs. (4.23) are positive, its value can be increased infinitely and without affecting the non-negativity requirements on $x_1, x_2, \ldots, x_m$. In such a case, there is no optimal value. However, if one or more of the coefficients of x_r in Eqs. (4.23) are negative, there will be a limit on the value that can be given to x_r, because, by increasing the value of x_r, the associated basic variable might become negative. Hence, the maximum value that can be assigned to x_r is the minimum value of x_r that will make one of the existent basic variables zero. This particular variable now becomes non-basic, giving place to x_r. In this way, the value of Z is increased.

Each iteration performed in the foregoing manner yields a new basic variable, with an increase in the value of the objective function. Ultimately, at least one set of basic variables gives an optimal solution to Z. This can be identified by the fact that the coefficients of all the non-basic variables in the objective function are positive, in which case no improvement in the value of Z is possible by making any of these non-basic variables assume a non-zero value. The coefficients of the non-basic variables in the objective function are generally called the *Simplex coefficients*.

All these arguments fail when the solution becomes degenerate during the iteration. In such a case, as shown in Section 4.5, the solution can be optimal even when all the Simplex coefficients in Eq. (4.24) are positive. Hence, the Simplex criterion is a *necessary* condition for optimality only when the basic solution is not degenerate. Further, when all the Simplex coefficients are positive, the solution is not only optimal but also unique. If some of the non-basic variables in Eq. (4.24) have zero coefficients, the solution can still be optimal; but, in this case, the problem will have more than one basic optimal solution.

EXERCISES

1. Precision optical parts are produced in a certain factory using two machines, A and B, and an optical grinder, G. The optical parts come in four types, namely, Regular, Super, Deluxe, and Super Deluxe. The processing time in hours per load required by each machine for each of the four types, the net profit per week per unit of each type, and the maximum time available per week on each machine are as tabulated here.

Type	*A*	*B*	*G*	*Profit*
Regular	2	2	1.5	\$ 6
Super	1	2	6.0	4
Deluxe	10	6	4.5	9
Super Deluxe	5	6	18.0	7
Max. Hours Available	50	36	81	

How many of each type of optical unit should the factory produce in order to make the maximum profit per week, and how much is this profit?
[*Ans.* Produce only Regulars, 18 per week; profit: \$ 108.]

2. A company has 1000 tonnes of ore B_1, 2000 tonnes of ore B_2, and 500 tonnes of ore B_3. Products A, B, and C can be produced from these ores. For 1 tonne of each of these products, the ore requirements are: A—5 tonnes of B_1, 10 tonnes of B_2, and 10 tonnes of B_3; B—5 tonnes of B_1, 8 tonnes of B_2, and 5 tonnes of B_3; C—10 tonnes of B_1, 5 tonnes of B_2, and none of B_3. The company makes a profit of \$ 100 on each tonne of A, \$ 200 on each tonne of B, and \$ 50 on each tonne of C. How many tonnes of each A, B, and C should the company produce to maximize its profit, and how much is this profit?
[*Ans.* A: 0; B: 100 tonnes; C: 50 tonnes; profit: \$ 22,500.]

3. Maximize $Z = 2x_1 + 4x_2 + x_3 + x_4$
subject to the constraints

$$x_1 + 3x_2 + x_4 \leqslant 4,$$

$$2x_1 + x_2 \leqslant 3,$$

$$x_2 + 4x_3 + x_4 \leqslant 3,$$

$$x_j \geqslant 0 \quad (j = 1, 2, 3, 4).$$

[*Ans.* $Z_{\max} = 6.5$.]

4. Maximize $Z = 2x_1 + 3x_2 + x_3 + 7x_4$

subject to the constraints

$$8x_1 + 3x_2 + 4x_3 + x_4 \leqslant 6,$$

$$2x_1 + 6x_2 + x_3 + 5x_4 \leqslant 3,$$

$$x_1 + 4x_2 + 5x_3 + 2x_4 \leqslant 7,$$

$$x_1, x_2, x_3, x_4, \text{ all} \geqslant 0.$$

[*Ans.* $Z_{\max} = 4.2$.]

5. Maximize $Z = x_1 + 4x_2 + 5x_3$
subject to the constraints

$$3x_1 + 6x_2 + 3x_3 \leqslant 22,$$

$$x_1 + 2x_2 + 3x_3 \leqslant 14,$$

$$3x_1 + 2x_2 \leqslant 14,$$

$$\text{all } x\text{'s} \geqslant 0.$$

[*Ans.* $Z_{\max} = 24\frac{2}{3}$.]

6. A small-scale industrialist produces four types of machine components, M_1, M_2, M_3, and M_4, made of steel and brass. The amounts of steel and brass required for each component and the number of man-weeks required to manufacture and assemble one unit of each component

	M_1	M_2	M_3	M_4	*Availability*
Steel	6	5	3	2	100 kg
Brass	3	4	9	2	75 kg
Man-weeks	1	2	1	2	20

are as tabulated here. The labour is restricted to 20 man-weeks, steel to 100 kg per week, and brass to 75 kg per week. The industrialist's profit on each unit of M_1, M_2, M_3, and M_4 is respectively \$ 6, \$ 4, \$ 7, and \$ 5.

How many of each component should he produce per week to optimize his profit, and how much is this profit? Note that the values given are the average values per week. If a fractional value appears in the answer, it should be interpreted as an average value.
[*Ans.* M_1: 14; M_2: 0; M_3: 10/3; M_4: 0; profit: \$ $113\frac{1}{3}$.]

7. Maximize $Z = -5x_1 + 10x_2 - 7x_3 + 3x_4$
subject to

$$x_1 + x_2 + 7x_3 + 2x_4 = 7/2,$$

$$-2x_1 - x_2 + 3x_3 + 3x_4 = 3/2,$$

$$2x_1 + 2x_2 + 8x_3 + x_4 = 4,$$

$$x_1, x_2, x_3, x_4, \text{ all} \geqslant 0.$$

[*Ans.* $Z_{max} = 18$, $x_1 = x_3 = 10$, $x_2 = 3/2$, $x_4 = 1$.]

8. Maximize $Z = 10x_1 - x_2 - 9x_3 - 8x_4$
subject to

$$2x_1 - x_2 - 3x_3 - x_4 = -2,$$

$$5x_1 - 2x_2 - 3x_4 = -5,$$

$$-7x_1 + 4x_2 - x_3 - 4x_4 \geqslant -1,$$

$$-3x_1 - 2x_2 - 5x_3 - 6x_4 \geqslant -10.$$

[*Ans.* $Z_{max} = -62/7$, $x_1 = 1/7$, $x_2 = 8/7$, $x_3 = 0$, $x_4 = 8/7$.]

9. Minimize $Z = -3x_1 + 6x_2$
subject to the constraints

$$x_1 + 3x_2 \geqslant -1,$$

$$2x_1 + x_2 \geqslant 4,$$

$$x_1 - x_2 \geqslant -1,$$

$$x_1 - 4x_2 \geqslant -13,$$

$$-4x_1 + x_2 \geqslant -23,$$

$$x_1, x_2 \geqslant 0.$$

[*Ans.* $Z_{min} = -33$, $x_1 = 5$, $x_2 = -3$.]

10. Maximize $Z = 2x_1 - 3x_2 + 5x_3$; $x_1, x_2, x_3 \geqslant 0$;
subject to the constraints

$$2x_1 - x_2 + 3x_3 \leqslant 4,$$

$$x_1 + x_2 \geqslant 6,$$

$$3x_1 + 2x_2 + 2x_3 \leqslant 7$$

and

(a)

$$-x_1 + 5x_2 + x_3 = 2$$

or

(b)

$$-x_1 + 5x_2 + x_3 = 16.$$

[*Ans.* No solution to (a) since it contradicts the other constraints; for (b), $Z = 12$, $(\frac{16}{21}, \frac{55}{21}, \frac{11}{3})$.]

11. In Exercise 10, determine the smallest value of the right-hand side term in the equality given by (a) to give a feasible solution.
[*Ans.* 5.]

12. Maximize $Z = 5x_1 + 6x_2 + 3x_3 + 2x_4 + 4x_5$
subject to the constraints

$$2x_1 + 3x_2 + x_3 + 5x_4 + 6x_5 \leqslant 100,$$

$$x_1 + 4x_2 + 4x_3 + 2x_4 + 2x_5 \leqslant 50,$$

$$0 \leqslant x_j \leqslant 6 \qquad (j = 1, 2, 3, 4, 5).$$

[*Ans.* $Z = 98$, (6, 6, 0, 4, 6).]

5

Two Phases of Simplex Method

5.1 INITIAL BASIC-FEASIBLE SOLUTION

In preparing the Simplex tableau, we assumed that a set of basic variables giving the feasible solution was available to us. When the constraint equations in a problem are few in number or involve slack variables, we can generally choose the initial basic variables fairly easily. When such a selection is not obvious, it is desirable to have a method that will provide an initial set of basic variables leading to the feasible solution. We shall now consider a procedure that will not only lead to such a solution, if it exists, but will also eliminate redundant equations among the constraint conditions, or terminate the iteration whenever a feasible solution does not exist.

Let the linear programming problem be available in the standard form, as represented by Eqs. (2.14), (2.15), and (2.16). Let all the b_r's ($r = 1, 2, \ldots, m$) be non-negative. If, in any equation, b_i happens to be negative, we shall multiply that equation by (-1) before the procedure is applied. This does not change the generality.

Introducing a set of non-negative variables called *artificial variables*, $x_{n+1}, x_{n+2}, \ldots, x_{n+m}$, one for each of the constraint equations, we get

$$\begin{aligned} a_{11}x_1 + a_{12}x_2 + \ldots + a_{1n}x_n + x_{n+1} &= b_1 \\ a_{21}x_1 + a_{22}x_2 + \ldots + a_{2n}x_n + x_{n+2} &= b_2 \\ &\vdots \\ a_{m1}x_1 + a_{m2}x_2 + \ldots + a_{mn}x_n + x_{n+m} &= b_m \end{aligned} \tag{5.1}$$

Let the sum of the artificial variables be denoted by

$$w = x_{n+1} + x_{n+2} + \ldots + x_{n+m}. \tag{5.2}$$

If the linear programming problem has a feasible solution, then w must be equal to zero, since all the artificial variables must be eliminated. Equation (5.2) may therefore be considered as the objective function in which w is to be minimized. If $w_{\min} = 0$, then a feasible solution exists; otherwise, no feasible solution exists. The function w is known as the *infeasibility form*.

Subtracting each equation in (5.1) from Eq. (5.2), we get

$$w = w_0 + d_1x_1 + d_2x_2 + \ldots + d_nx_n, \tag{5.3}$$

where

$$d_1 = -(a_{11} + a_{21} + \ldots + a_{m1}) = -\sum_{i=1}^{m} a_{i1}$$

$$d_2 = -(a_{12} + a_{22} + \ldots + a_{m2}) = -\sum_{i=1}^{m} a_{i2}$$

$$\vdots$$

$$d_n = -(a_{1n} + a_{2n} + \ldots + a_{mn}) = -\sum_{i=1}^{m} a_{in}$$

$$w_0 = b_1 + b_2 + \ldots + b_m = -\sum_{i=1}^{m} b_i.$$

Now, we can treat Eqs. (5.1) as a linear programming problem, with Eq. (5.3) as the objective function. This new problem has $(n + m)$ variables. Since all the artificial variables, $x_{n+1}, x_{n+2}, \ldots, x_{n+m}$, are positive, with their coefficients unity, they can form the initial basis. In Eqs. (5.1), there are as many artificial variables as the number of constraints. It should be observed that Eq. (5.3) does not say anything new. If the constraints are consistent, then there are suitable slack variables to make the constraints assume the form of equations as given by Eq. (2.16). Equation (5.3) is nothing but the sum of these equalities with the right-hand side quantity $\Sigma\, b_i$ shifted to the left-hand side so that w becomes equal to zero. If w is not zero, then the constraints are inconsistent.

It is possible to reduce the number of artificial variables by taking advantage of those slack variables that have positive coefficients. However, we should be careful with regard to two aspects of such a case: first, the infeasibility form w should consist of only the artificial variables; second, we should subtract from the infeasibility form only those constraint equations that contain the artificial variables, not those containing slack variables, with positive coefficients, which were used in place of the artificial variables.

We shall demonstrate the foregoing procedure through three examples: first, a well-behaved case (Problem 5.1); second, one that involves inconsistent constraint equations (Problem 5.2); and third, a problem containing a redundant constraint equation (Problem 5.3).

Problem 5.1

Maximize $Z = 2x_1 - x_2 + x_3$

subject to the constraints

$$x_1 + x_2 - 3x_3 \leqslant 8,$$

$$4x_1 - x_2 + x_3 \geqslant 2,$$
$$2x_1 + 3x_2 - x_3 \geqslant 4,$$
$$\text{all } x_r\text{'s} \geqslant 0.$$

Putting this problem in the standard form, without making it a minimization problem, we get:

$$\text{Maximize } Z = 2x_1 - x_2 + x_3$$

subject to

$$x_1 + x_2 - 3x_3 + x_4 = 8,$$
$$4x_1 - x_2 + x_3 - x_5 = 2,$$
$$2x_1 + 3x_2 - x_3 - x_6 = 4,$$
$$\text{all } x_r\text{'s} \geqslant 0.$$

Of the three slack variables x_4, x_5, and x_6, only x_4 appears with a $(+1)$ coefficient. Hence, we shall introduce two artificial variables, x_7 and x_8, and obtain

$$x_1 + x_2 - 3x_3 + x_4 = 8,$$
$$4x_1 - x_2 + x_3 - x_5 + x_7 = 2,$$
$$2x_1 + 3x_2 - x_3 - x_6 + x_8 = 4,$$
$$\text{all } x_r\text{'s} \geqslant 0.$$

The infeasibility form is

$$w = x_7 + x_8.$$

Subtracting only the second and third equations from w (since we have made use of the slack variable x_4 as one of the artificial variables), we get

$$w = 6 - 6x_1 - 2x_2 + x_5 + x_6.$$

We shall take this as the new objective function and proceed to minimize w in Tableaux I–III.

TABLEAU I

Basis	x_1	x_2	x_3	x_4	x_5	x_6	x_7	x_8	w	
x_4	1	1	−3	1						8
x_7	4	−1	1		−1		1			2
x_8	2	3	−1			−1		1		4
	6 ↑	2			−1	−1			1	6

Since the new problem is one of minimization, we select the most positive coefficient in the last row. In Tableau II, the new variable x_1 replaces the variable x_7.

TABLEAU II

Basis	x_1	x_2	x_3	x_4	x_5	x_6	x_7	x_8	w	
x_4		5/4	−13/4	1	1/4		−1/4			15/2
x_1	1	−1/4	1/4		−1/4		1/4			1/2
x_8		7/2	−3/2		1/2	−1	−1/2	1		3
		7/2 ↑	−3/2		1/2	−1	−3/2		1	3

Tableau III is obtained by replacing the variable x_8 by the new variable x_2.

TABLEAU III

Basis	x_1	x_2	x_3	x_4	x_5	x_6	x_7	x_8	w	
x_4			−19/7	1	1/14	5/14	−1/14	−5/14		45/7
x_1	1		1/7		−3/14	−1/14	3/14	1/14		5/7
x_2		1	−3/7		1/7	−2/7	−1/7	2/7		6/7
			0		0	0	−1	−1	1	0

The optimal solution is zero; hence, an initial basic set of variables that will lead to a feasible solution is (x_1, x_2, x_4). Having obtained this set, we can proceed to solve the original linear programming problem.

Problem 5.2

Minimize $Z = 8x_1 + 10x_2 + 12x_3 + 11x_4$

subject to

$$5x_2 + 4x_3 + 4x_4 = 5,$$

$$6x_1 + 4x_2 + 3x_3 + 2x_4 = 3,$$

$$4x_1 + x_2 + 3x_3 + 4x_4 = 2,$$

$$x_1, x_2, x_3, x_4, \text{ all} \geqslant 0.$$

We introduce into these equations the artificial variables x_5, x_6, and

x_7, and obtain

$$5x_2 + 4x_3 + 4x_4 + x_5 = 5,$$

$$6x_1 + 4x_2 + 3x_3 + 2x_4 + x_6 = 3,$$

$$4x_1 + x_2 + 3x_3 + 4x_4 + x_7 = 2.$$

The infeasibility form is

$$w = 10 - 10x_1 - 10x_2 - 10x_3 - 10x_4.$$

With this as our new objective function, we proceed to minimize w in Tableaux I–III.

TABLEAU I

Basis	x_1	x_2	x_3	x_4	x_5	x_6	x_7	w	
x_5		5	4	4	1				5
x_6	6	4	3	2		1			3
x_7	4	1	3	4			1		2
	10 ↑	10	10	10				1	10

The objective function asserts that we select any of the variables x_1, x_2, x_3, x_4 as the new basic variable. If we select x_1, there will be two contenders, namely, x_6 and x_7, for the replaced variable which will lead us to a degenerate case (as discussed in Section 4.5). Nevertheless, let us choose x_1 to note the result. Of the two competitors x_6 and x_7, we shall arbitrarily select x_7 as the replaced variable. This gives us Tableau II.

TABLEAU II

Basis	x_1	x_2	x_3	x_4	x_5	x_6	x_7	w	
x_5		5	4	4	1				5
x_6		5/2	−3/2	−4		1	−3/2		0
x_1	1	1/4	3/4	1			1/4		1/2
	0	15/2 ↑	5/2	0			−5/2	1	5

As Tableau II shows, we have obtained a degenerate case. We can avoid this situation by selecting x_2, x_3, or x_4 as the new basic variable. We proceed formally and select x_2 as the next basic variable that replaces x_6, giving us Tableau III.

TABLEAU III

Basis	x_1	x_2	x_3	x_4	x_5	x_6	x_7	w	
x_5			7	12	1	-2	-3		5
x_2		1	$-3/5$	$-8/5$		$2/5$	$-3/5$		0
x_1	1		$9/10$	$7/5$		$-1/10$	$2/5$		$1/2$
	0	0	7	12 ↑		-3	0	1	5

With x_4 as the new basic variable, we get Tableau IV.

TABLEAU IV

Basis	x_1	x_2	x_3	x_4	x_5	x_6	x_7	w	
x_5	$-60/7$		$-5/7$		1	$-8/7$	-5		$5/7$
x_2	$8/7$	1	$3/7$			$5/7$	$-37/35$		$4/7$
x_4	$5/7$		$9/14$	1		$-1/14$	$-2/7$		$5/14$
	$-60/7$	0	$-5/7$	0		$-15/7$	$-24/7$	1	$5/7$

The coefficients of all the variables in the last row are negative, but $w \neq 0$. Hence, the problem does not have a feasible solution.

That the problem is infeasible can be easily deduced. Adding all the constraint equations, we get

$$x_1 + x_2 + x_3 = 1.$$

Hence, x_1, x_2, x_3, x_4, all $\leqslant 1$. From the first constraint equation, the only possible answer is $x_2 = 1$, and $x_3 = x_4 \equiv 0$. This cannot satisfy either the second or the third equation.

Problem 5.3 Solve the following problem which involves a redundant constraint equation:

Determine $x_1, x_2, x_3, x_4 \geqslant 0$

so as to

$$\text{minimize } Z = 3x_1 + 2x_2 + x_3 + 1.5x_4$$

subject to

$$0.4x_1 + 0.2x_2 + 0.5x_3 + 0.8x_4 = 0.3,$$
$$0.6x_1 + 0.8x_2 + 0.5x_3 + 0.2x_4 = 0.7,$$
$$x_1 + x_2 + x_3 + x_4 = 1.0.$$

The last constraint equation is obviously redundant since it can be obtained by adding the first two. Nevertheless, we include it in the set to note the result. We shall introduce into these three equations three artificial variables, x_5, x_6, and x_7, and obtain

$$0.4x_1 + 0.2x_2 + 0.5x_3 + 0.8x_4 + x_5 = 0.3,$$

$$0.6x_1 + 0.8x_2 + 0.5x_3 + 0.2x_4 + x_6 = 0.7,$$

$$x_1 + x_2 + x_3 + x_4 + x_7 = 1.0.$$

The infeasibility form is

$$w = x_5 + x_6 + x_7 = 2 - 2x_1 - 2x_2 - 2x_3 - 2x_4.$$

With this as our new objective function, we proceed to minimize w in Tableaux I–III.

TABLEAU I

Basis	x_1	x_2	x_3	x_4	x_5	x_6	x_7	w	
x_5	***0.4***	0.2	0.5	0.8	1				0.3
x_6	0.6	0.8	0.5	0.2		1			0.7
x_7	1	1	1	1			1		1
	2	2	2	2				1	2
	↑								

TABLEAU II

Basis	x_1	x_2	x_3	x_4	x_5	x_6	x_7	w	
x_1	1	1/2	5/4	2	1/4				3/4
x_6		*1/2*	−1/4	−1	−3/20	1			1/4
x_7		1/2	−1/4	−1	−1/4		1		1/4
	0	1	−1/2	−2	−1/2			1	1/2
		↑							

TABLEAU III

Basis	x_1	x_2	x_3	x_4	x_5	x_6	x_7	w	
x_1	1		3/2	3	1	1			1
x_2		1	−1/2	−2	−3/10	2			1/2
x_7					−1/10	−1	1		0
	0	0	0	0	−17/10	−2		1	0

Tableau III shows that we have obtained an optimal solution to the new problem. With x_1, x_2, and x_7 as the basic variables, we get the solutions

$$x_1 = 1, \qquad x_2 = 1/2, \qquad x_7 = 0.$$

This means that x_7 is no longer involved in the calculations. In essence, this is equivalent to saying that the equation that brought in the artificial variable x_7 is redundant.

5.2 TWO-PHASE PROCESS

The procedure adopted in Section 5.1 provides us with an initial basic-feasible solution which can be used as the basis for the first tableau in the Simplex method. These two procedures—the first dealing with the determination of an initial set of basic variables giving a feasible solution, and the second with the Simplex method—are together called the two-phase process.

We must be careful to observe that the objective function for each of the two phases is different: whereas in Phase I the objective function is made up of only the artificial variables (of the new problem), Phase II uses the original objective function (of the original problem). The two procedures can be performed successively. In the tableaux, we enter both the original objective function and the infeasibility form. In Phase I, the Simplex criterion is applied only to the coefficients of the infeasibility form. However, in each cycle, when one basis is changed to another, the coefficients of the original objective function are also subjected to the same "transformations". In this way, when we have obtained the basic-feasible solution at the end of Phase I, the objective function will contain only the non-basic variables in a form suitable to the beginning of Phase II. These steps are shown in Problem 5.4.

Problem 5.4 Solve the following linear programming problem, using the two phases of the Simplex method:

$$\text{Minimize } Z = 2x_1 + x_2$$

subject to

$$5x_1 + 10x_2 - x_3 = 8,$$

$$x_1 + x_2 + x_4 = 1,$$

$$x_1, x_2, x_3 \; x_4; \text{ all} \geqslant 0.$$

Since x_4 is appearing with a (+1) coefficient, we shall introduce only one artificial variable, x_5. We then get

$$5x_1 + 10x_2 - x_3 + x_5 = 8,$$

$$x_1 + x_2 + x_4 = 1.$$

The infeasibility form is

$$w \equiv x_5 = 8 - 5x_1 - 10x_2 + x_3.$$

With this as our new objective function, we proceed to Tableau I.

TABLEAU I (Phase I, Cycle 1)

Basis	x_1	x_2	x_3	x_4	x_5	Z	w	
x_5		*10*	−1		1			8
x_4	1	1		1				1
Z form	−2	−1				1		0
w form	5	10 ↑	−1				1	8

Bringing in x_2 in place of x_5, we get Tableau II. The simplex criterion shows that we have obtained the optimal solution with x_2 and x_4 as the basic variables. So, in this particular example, the end of Phase I gives the optimal result.

TABLEAU II (Phases I & II, Cycle 2)

Basis	x_1	x_2	x_3	x_4	x_5	Z	w	
x_2	1/2	1	−1/10		1/10			4/5
x_4	1/2		1/10	1	−1/10			1/5
Z form	−3/2					1		4/5
w form					−1		1	0

5.3 *M*-METHOD

The discussions on the two-phase process (Section 5.2) and the use of artificial variables (Section 5.1) have indicated how an initial basic-feasible solution may be obtained. A variant of this method, generally known as the *M-method*, is frequently used to obtain the initial basic-feasible solution. Consider the following problem.

Problem 5.5

Minimize $Z = 2x_1 + x_2$

subject to the constraints

$$3x_1 + x_2 \geqslant 3,$$

$$4x_1 + 3x_2 \geqslant 6,$$

$$x_1 + 2x_2 \geqslant 2,$$

$$x_1, x_2 \geqslant 0.$$

Introducing the slack variables x_3, x_4, and x_5, we find that the constraints become

$$3x_1 + x_2 - x_3 = 3,$$

$$4x_1 + 3x_2 - x_4 = 6,$$

$$x_1 + 2x_2 - x_5 = 2,$$

$$\text{all } x_i\text{'s} \geqslant 0.$$

Let us introduce the artificial variables x_6, x_7, and x_8 such that the constraints become

$$3x_1 + x_2 - x_3 + x_6 = 3,$$

$$4x_1 + 3x_2 - x_4 + x_7 = 6,$$

$$x_1 + 2x_2 - x_5 + x_8 = 2,$$

$$x_1, x_2, \ldots, x_8 \geqslant 0.$$

An immediate feasible solution is $x_6 = 3$, $x_7 = 6$, and $x_8 = 2$. The introduction of the artificial variables has changed the problem but, in the process of arriving at the optimal solution, if the artificial variables do not occur or appear with zero value, the optimal solution will correspond to the original problem. In order to ensure this, the objective function is transformed into

$$Z_m = 2x_1 + x_2 + M(x_6 + x_7 + x_8),$$

where M is a large number. M should be larger than any number appearing during optimization. If the original problem has a solution, then the modified problem also has a solution with all the artificial variables having zero value. The value of Z_m with all the artificial variables zero will be lesser than any other value of Z_m in which the artificial variables are non-zero positive quantities. Hence, if the original problem has a solution, then that solution may be obtained through the solution of the modified problem, and no artificial variable appears in the final solution. If, on the other hand, the optimal solution to the modified solution appears with positive non-zero values for some of or all the artificial variables, then the constraints to the original problem are inconsistent.

If the problem involves the maximization of an objective function, then the modified problem has an objective function in which the artificial variables appear with $-M$ factors. The modified problem can be solved as shown in Tableaux I to V.

TABLEAU I

Basis	x_1	x_2	x_3	x_4	x_5	x_6	x_7	x_8	
x_6	3	1	-1			1			3
x_7	4	3		-1			1		6
x_8	1	2			-1			1	2
	-2	-1				$-M$	$-M$	$-M$	0

To make x_6, x_7, and x_8 basic variables, we have to make their coefficients zero in the last row. The resulting tableau is shown as Tableau II.

TABLEAU II

Basis	x_1	x_2	x_3	x_4	x_5	x_6	x_7	x_8	
x_6	3	1	-1			1			3
x_7	4	3		-1			1		6
x_8	1	2			-1			1	2
	-2	-1				0	0	0	
	$+8M$	$+6M$	$-M$	$-M$	$-M$				$11M$

In Tableau II, the terms in the row corresponding to the objective function are entered in two rows for convenience; the coefficient of x_1 is $(-2+8M)$ and that of x_2 is $(-1+6M)$; and the numbers with M-factors appear in a separate row. Since $(8M-2)$ is greater than all the other coefficients, x_1 replaces x_6 as the basis in Tableau III. In this process x_6 becomes non-basic and can therefore be ignored, since it has served its purpose. As long as it remains non-basic, its value is zero in the final optimal solution. On the other hand, if it remains as a basic variable even when it has a zero value, it cannot be ignored. This is an important point to observe.

TABLEAU III

Basis	x_1	x_2	x_3	x_4	x_5	x_6	x_7	x_8	
x_1	1	1/3	$-1/3$						1
x_7	0	5/3	4/3	-1			1		2
x_8	0	5/3	1/3		-1			1	1
	0	$-1/3$	$-2/3$				0	0	2
		$\frac{10}{3}M$	$\frac{5}{3}M$	$-M$	$-M$				$3M$

Since x_8 has now become non-basic, it is omitted in Tableau IV.

TABLEAU IV

Basis	x_1	x_2	x_3	x_4	x_5	x_7	
x_1	1	0	$-2/5$		$1/5$		$4/5$
x_7	0	0	1	-1	*1*		1
x_2	0	1	$1/5$		$-3/5$	$3/5$	$3/5$
	0	0	$-3/5$		$-1/5$	0	$11/5$
			M	$-M$	M		M

In Tableau V, x_5 replaces x_7 in the basis.

TABLEAU V

Basis	x_1	x_2	x_3	x_4	x_5	
x_1	1	0	$-3/5$	$1/5$	0	$3/5$
x_5	0	0	1	-1	1	1
x_2	0	1	$4/5$	$-3/5$	0	$6/5$
	0	0	$-2/5$	$-1/5$	0	$12/5$

Tableau V gives the optimal solution, with

$x_1 = 3/5, \quad x_2 = 6/5, \quad x_3 = 0, \quad x_4 = 0, \quad x_5 = 1,$

$Z = 12/5.$

EXERCISES

1. Use the infeasibility form to determine the initial basic-feasible solutions to the following sets of constraints:

(a)

$$3x_1 + 6x_2 + 3x_3 \leqslant 22,$$

$$x_1 + 2x_2 + 3x_3 \leqslant 14,$$

$$3x_1 + 2x_2 \leqslant 14.$$

(b)

$$8x_1 + 3x_2 + 4x_3 + x_4 \leqslant 6,$$

$$2x_1 + 6x_2 + x_3 + 5x_4 \leqslant 3,$$

$$x_1 + 4x_2 + 5x_3 + 2x_4 \leqslant 7.$$

2. Use the infeasibility form to test the following sets of equations for consistency and redundancy:

(a)

$$2x + 4y + 3z + u = 15,$$

$$3x + 7y + 2u = 16,$$

$$5x + 3y + 2z + 3u = 21.$$

[*Ans.* Consistent.]

(b)

$$4x + 8y + 6z + v = 18,$$

$$2x + 3y + 4z = 8,$$

$$7x + (27/x)y + 11z + (3/2)v = 31.$$

[*Ans.* Inconsistent.]

(c)

$$x_1 + x_2 + x_3 + x_4 = 10,$$

$$2x_1 + x_2 + 2x_3 = 10,$$

$$x_1 + 2x_2 + x_3 + 3x_4 = 20.$$

[*Ans.* Redundant.]

3. Solve Problem 4 (Chapter 4) using the two-phase process.

4. Solve Problem 5 (Chapter 4) using the two-phase process.

5. A special machine component requires processing on two different lathes, A and B, and also on a grinder, G. These components come in four different sizes, C_1, C_2, C_3, and C_4. The processing time (in hours) required by each machine for each of the four types of component and also the net profit for each type are tabulated here. The maximum time available per week on each machine is also indicated. How many of each component should be processed in order to maximize the profit?

Type	*Lathe A*	*Lathe B*	*Grinder G*	*Profit*
C_1	10	6	4.5	Rs 90
C_2	5	6	18.0	Rs 70
C_3	2	2	1.5	Rs 60
C_4	1	2	6.0	Rs 40
Max. Hours Available	50	36	81	

[*Ans.* Process only C_4 (18 components); profit: Rs 1080.]

6. Maximize $Z = x_1 + x_2 + x_3$, $x_j \geqslant 0\,(j = 1, 2, 3)$, subject to the constraints

$$3x_1 - x_2 - x_3 \leqslant 1,$$

$$-x_1 + 3x_2 - x_3 \leqslant 1,$$

$$x_1 - x_2 + 3x_3 \leqslant 1.$$

[*Ans.* $Z = 5/3$, $(2/3, 2/3, 1/3)$.]

7. What will be the result in Problem 6 if the second constraint is omitted?
[*Ans.* $Z = -\infty$.]

8. Use the M-method to

$$\text{maximize } Z = 2x_1 + 7x_2$$

subject to the constraints

$$2x_1 + 3x_2 \leqslant 10,$$

$$x_1 + 3x_2 \leqslant 5,$$

$$x_1, x_2 \geqslant 0.$$

[*Ans.* $Z = 70/3$, $x_2 = 10/3$.]

9. Use the M-method to solve the following problem:

$$\text{Minimize } Z = 3y_3 + y_4 + 2y_5$$

subject to

$$y_3 + y_4 - 2y_5 - y_1 = 1,$$

$$y_3 - 2y_4 + y_5 - y_2 = -1,$$

$$\text{all } y\text{'s} \geqslant 0.$$

[*Ans.* $Z = 5/3$, $y_4 = 2/3$, $y_3 = 1/3$.]

6

Duality Theorem

6.1 CONSTRUCTION OF DUAL

An important characteristic of the linear programming model is that it possesses a dual which has certain similarities. Every minimization problem has a corresponding maximization problem, both involving the same data and with closely-connected optimal solutions. The two problems are said to be *duals* of each other, and to distinguish between them one is called the *primal* and the other the *dual*.

To see how a dual problem is formulated when a primal is given, let us consider the following linear programming problem:

$$\text{Minimize } Z = c_1x_1 + c_2x_2 + \ldots + c_nx_n$$

subject to the constraints

$$a_{11}x_1 + a_{12}x_2 + \ldots + a_{1n}x_n \geqslant b_1$$

$$a_{21}x_1 + a_{22}x_2 + \ldots + a_{2n}x_n \geqslant b_2$$

$$\vdots$$

$$a_{m1}x_1 + a_{m2}x_2 + \ldots + a_{mn}x_n \geqslant b_m$$

$$x_1 \geqslant 0, x_2 \geqslant 0, \ldots, x_n \geqslant 0.$$

To construct the dual, we adopt the following steps.

(i) For a primal with n variables and m constraint conditions, the dual will have m variables with n constraint conditions.

(ii) The greater-than or equal-to signs of the primal constraint conditions become less-than or equal-to signs in the dual constraint conditions.

(iii) The minimization problem in the primal becomes a maximization problem in the dual.

(iv) The constants $c_1, c_2, \ldots, c_n$ in the objective function of the primal become $b_1, b_2, \ldots, b_m$ in the objective function of the dual.

(v) The constants $b_1, b_2, \ldots, b_m$ in the constraints of the primal become $c_1, c_2, \ldots, c_n$ in the objective function of the dual.

(vi) In both the primal and the dual, the variables are non-negative.

Using these six guidelines, we obtain the dual of the primal:

$$\text{Maximize } w = b_1y_1 + b_2y_2 + \ldots + b_my_m$$

subject to the constraints

$$a_{11}y_1 + a_{21}y_2 + a_{31}y_3 + \ldots + a_{m1}y_m \leqslant c_1$$

$$a_{12}y_1 + a_{22}y_2 + a_{32}y_3 + \ldots + a_{m2}y_m \leqslant c_2$$

$$\vdots$$

$$a_{1n}y_1 + a_{2n}y_2 + a_{3n}y_3 + \ldots + a_{mn}y_m \leqslant c_n$$

$$y_1 \geqslant 0, y_2 \geqslant 0, \ldots, y_m \geqslant 0.$$

The dual of the dual is the primal.

As an example of duality, let us consider Problem 6.1.

Problem 6.1 Construct the dual to the primal given as:

$$\text{Minimize } Z = 2x_1 + 9x_2 + 5x_3$$

subject to the constraints

$$x_1 + x_2 - x_3 \geqslant 1,$$

$$-2x_1 + x_3 \geqslant 2,$$

$$x_1 \geqslant 0, x_2 \geqslant 0, x_3 \geqslant 0.$$

The dual problem is:

$$\text{Maximize } w = y_1 + 2y_2$$

subject to the constraints

$$y_1 - 2y_2 \leqslant 2,$$

$$y_1 \leqslant 9,$$

$$-y_1 + y_2 \leqslant 5,$$

$$y_1 \geqslant 0, y_2 \geqslant 0.$$

Note The construction of a dual according to Steps (i) to (vi) is feasible for a primal stated as a minimization problem with greater-than or equal-to constraint conditions. If some of the constraints in the primal have less-than or equal-to signs, these should be converted into greater-than or equal-to signs by multiplying both sides of these equations by (−1). Further, if the primal problem is concerned with maximization, it needs to be changed to a minimization problem by multiplying the objective function by (−1) before its dual can be formed. As an example, let us consider Problem 6.2.

Problem 6.2 Construct the dual of the primal problem given as:

$$\text{Maximize } Z = 4x_1 + 18x_2 + 10x_3$$

subject to the constraints

$$x_1 + x_2 - x_3 \geqslant 2,$$
$$-2x_1 + x_3 \leqslant 1,$$
$$x_1 \geqslant 0,\ x_2 \geqslant 0,\ x_3 \geqslant 0.$$

The maximization problem is converted into a minimization problem and the less-than sign in the second constraint condition is changed to a greater-than sign. Now, the primal is:

$$\text{Minimize } Z^* = (-Z) = -4x_1 - 18x_2 - 10x_3$$

subject to the constraints

$$x_1 + x_2 - x_3 \geqslant 2,$$
$$2x_1 - x_3 \geqslant -1,$$
$$x_1 \geqslant 0,\ x_2 \geqslant 0,\ x_3 \geqslant 0.$$

The dual of this primal is:

$$\text{Maximize } w = 2y_1 - y_2$$

subject to the constraints

$$y_1 + 2y_2 \leqslant -4,$$
$$y_1 \leqslant -18,$$
$$-y_1 - y_2 \leqslant -10,$$
$$y_1 \geqslant 0,\ y_2 \geqslant 0.$$

6.2 EQUALITY CONSTRAINTS

We have seen how a dual of a given primal can be formulated when the constraint conditions in the primal are in the form of inequalities. If some of the constraints in the primal are equalities, then some of the admissible variables in the dual become unrestricted in sign as will now be shown.

Let k number of constraints in the primal appear with equality signs as

$$\begin{aligned} a_{11}x_1 + a_{12}x_2 + \ldots + a_{1n}x_n &= b_1 \\ a_{21}x_1 + a_{22}x_2 + \ldots + a_{2n}x_n &= b_2 \\ &\vdots \\ a_{k1}x_1 + a_{k2}x_2 + \ldots + a_{kn}x_n &= b_k \end{aligned} \tag{6.1}$$

In order to apply Steps (i) to (vi), Section 6.1, in the formulation of a dual, we can express each equation in (6.1) as two inequalities. For example, if we have $x = 2$, this equality can be written as $x \geqslant 2$ and

$x \leqslant 2$, i.e., $x \geqslant 2$ and $-x \geqslant -2$. In the same manner, Eq. (6.1) can be written as

$$\begin{aligned} a_{11}x_1 + a_{12}x_2 + \ldots + a_{1n}x_n &\geqslant b_1 \\ -a_{11}x_1 - a_{12}x_2 - \ldots - a_{1n}x_n &\geqslant -b_1 \\ a_{21}x_1 + a_{22}x_2 + \ldots + a_{2n}x_n &\geqslant b_2 \\ -a_{21}x_1 - a_{22}x_2 - \ldots - a_{2n}x_n &\geqslant -b_2 \\ &\vdots \\ a_{k1}x_1 + a_{k2}x_2 + \ldots + a_{kn}x_n &\geqslant b_k \\ -a_{k1}x_1 - a_{k2}x_2 - \ldots - a_{kn}x_n &\geqslant -b_k \end{aligned} \tag{6.2}$$

In addition to constraints (6.2), we have in the primal other $(m - k)$ constraint inequalities in the form

$$\begin{aligned} a_{k+1,\,1}x_1 + a_{k+1,\,2}x_2 + \ldots + a_{k+1,\,n}x_n &\geqslant b_{k+1} \\ &\vdots \\ a_{m1}x_1 + a_{m2}x_2 + \ldots + a_{mn}x_n &\geqslant b_m \end{aligned} \tag{6.3}$$

Now, we can form a dual, assuming each inequality condition in (6.2) and (6.3) is separate. For example, using $s_1, s_1', s_2, s_2', \ldots, s_k, s_k'$ as the dual variables for constraints (6.2), and $y_{k+1}, y_{k+2}, \ldots, y_m$ as the dual variables for constraints (6.3), we get the first constraint condition in the dual as

$$(a_{11}s_1 - a_{11}s_1') + (a_{21}s_2 - a_{21}s_2') + \ldots + (a_{k1}s_k - a_{k1}s_k') + a_{k+1,\,1}y_{k+1} + \ldots + a_{m1}y_m \leqslant c_1,$$

and apply similar steps for the remaining constraints. As before, all the admissible variables here are non-negative, i.e.,

$$s_1 \geqslant 0,\ s_1' \geqslant 0,\ s_2 \geqslant 0,\ s_2' \geqslant 0,\ \ldots,\ y_m \geqslant 0. \tag{6.4}$$

When the constraints are regrouped they appear as

$$a_{11}(s_1 - s_1') + a_{21}(s_2 - s_2') + \ldots + a_{k1}(s_k - s_k') + a_{k+1,\,1}y_{k+1} + \ldots + a_{m1}y_m \leqslant c_1.$$

Let

$$(s_1 - s_1') = y_1,\ (s_2 - s_2') = y_2,\ \ldots . \tag{6.5}$$

Then the first constraint takes the form

$$a_{11}y_1 + a_{21}y_2 + \ldots + a_{m1}y_m \geqslant c_1. \tag{6.6}$$

Similar steps may be applied for the remaining constraints. Hence, even when some of the constraints in the primal appear with equality signs, we can formulate a dual by following the foregoing steps, but with an important difference in respect of the non-negativity character of the variables. We observe that whereas, according to constraints (6.4), $s_1, s_1', s_2, s_2', \ldots, y_{k+1}, \ldots, y_m$ are all non-negative, the values of $y_1, y_2, \ldots, y_k$ as defined in Eq. (6.5), i.e., $y_1 = s_1 - s_1'$, $y_2 = s_2 - s_2', \ldots,$ need not be non-negative, because the differences $s_1 - s_1', s_2 - s_2', \ldots$ can be either positive or negative. Hence, the k number of admissible variables appearing in the dual corresponding to the k number of constraint equalities of the primal are unrestricted in sign. The dual for the present case therefore appears as:

$$\text{Maximize } w = b_1y_1 + b_2y_2 + \ldots + b_my_m$$

subject to the constraints

$$a_{11}y_1 + a_{21}y_2 + \ldots + a_{m1}y_m \leqslant c_1$$

$$a_{12}y_1 + a_{22}y_2 + \ldots + a_{m2}y_m \leqslant c_2$$

$$\vdots$$

$$a_{1n}y_1 + a_{2n}y_2 + \ldots + a_{mn}y_m \leqslant c_n$$

where $y_1, y_2, \ldots, y_k$ are *unrestricted in sign* and $y_{k+1}, \ldots, y_m$ are *non-negative.*

As an example, consider Problem 6.3.

Problem 6.3 Construct the dual of the linear programming problem:

$$\text{Minimize } Z = -6x_1 - 8x_2 + 10x_3$$

subject to

$$3x_1 + x_2 - x_3 = 5,$$

$$2x_1 + 4x_2 - x_3 = 8,$$

$$x_1 \geqslant 0, x_2 \geqslant 0, x_3 \geqslant 0.$$

The dual problem is:

$$\text{Maximize } w = 5y_1 + 8y_2$$

subject to the constraints

$$3y_1 + 2y_2 \leqslant -6,$$

$$y_1 + 4y_2 \leqslant -8,$$

$$-y_1 - y_2 \leqslant 10,$$

where y_1 and y_2 are unrestricted in sign.

6.3 DUALITY THEOREM

Let the primal problem involving m inequalities in n structural variables (or decision variables) be:

$$\text{Minimize } Z = c_1x_1 + c_2x_2 + \ldots + c_nx_n$$

subject to the constraints

$$\begin{aligned} &a_{11}x_1 + a_{12}x_2 + \ldots + a_{1n}x_n \geqslant b_1 \\ &a_{21}x_1 + a_{22}x_2 + \ldots + a_{2n}x_n \geqslant b_2 \\ &\vdots \\ &a_{m1}x_1 + a_{m2}x_2 + \ldots + a_{mn}x_n \geqslant b_m \\ &x_1 \geqslant 0, x_2 \geqslant 0, \ldots, x_n \geqslant 0. \end{aligned} \tag{6.7}$$

The dual to this problem involves *m* structural variables and *n* inequalities and appears as:

$$\text{Maximize } w = b_1y_1 + b_2y_2 + \ldots + b_my_m$$

subject to the constraints

$$\begin{aligned} &a_{11}y_1 + a_{21}y_2 + \ldots + a_{m1}y_m \leqslant c_1 \\ &a_{12}y_1 + a_{22}y_2 + \ldots + a_{m2}y_m \leqslant c_2 \\ &\vdots \\ &a_{1n}y_1 + a_{2n}y_2 + \ldots + a_{mn}y_m \leqslant c_n \\ &y_1 \geqslant 0, y_2 \geqslant 0, \ldots, y_n \geqslant 0. \end{aligned} \tag{6.8}$$

Changing the inequalities to equalities, we can restate the two problems (including *m* slack variables $x_1', x_2', \ldots, x_m'$ in the primal and *n* slack variables $y_1', y_2', \ldots, y_n'$ in the dual) as follows.

Primal

$$\text{Minimize } Z = c_1x_1 + c_2x_2 + \ldots + c_nx_n$$

subject to

$$\begin{aligned} &a_{11}x_1 + a_{12}x_2 + \ldots + a_{1n}x_n - x_1' = b_1 \\ &a_{21}x_1 + a_{22}x_2 + \ldots + a_{2n}x_n + x_2' = b_2 \\ &\vdots \\ &a_{m1}x_1 + a_{m2}x_2 + \ldots + a_{mn}x_n - x_m' = b_m \\ &x_1 \geqslant 0, x_2 \geqslant 0, \ldots, x_n \geqslant 0; x_1' \geqslant 0, \ldots, x_m' \geqslant 0. \end{aligned} \tag{6.9}$$

Dual

$$\text{Maximize } w = b_1y_1 + b_2y_2 + \ldots + b_my_m$$

subject to

$$
\begin{aligned}
&a_{11}y_1 + a_{21}y_2 + \ldots + a_{m1}y_m + y_1' = c_1\\
&a_{12}y_1 + a_{22}y_2 + \ldots + a_{m2}y_m + y_2' = c_2\\
&\vdots\\
&a_{1n}y_1 + a_{2n}y_2 + \ldots + a_{mn}y_m + y_n' = c_n\\
&y_1 \geqslant 0,\ y_2 \geqslant 0, \ldots, y_m \geqslant 0;\ y_1' \geqslant 0, \ldots, y_n \geqslant 0.
\end{aligned}
\tag{6.10}
$$

As per the duality theorem, in the optimal basic solutions of Eqs. (6.9) and (6.10),

(i) the values of x_i and x_j' in the optimal basis are numerically equal to the Simplex coefficients of y_i' and y_j, respectively; and the values of y_i and y_j' in the optimal basis are numerically equal to the Simplex coefficients of x_i' and x_j, respectively; and

(ii) $Z_{\min} = w_{\max}$.

The duality theorem will be proved in Chapter 10. Here we shall demonstrate the validity of the theorem by working out a few problems. Before we do this, some important properties of the duality relations may be easily derived from the definitions of primal and dual problems. Let us consider constraints (6.7). Multiplying the first of these constraints by y_1, the second by y_2, and so on, and adding all the results, we get

$$
\begin{aligned}
&a_{11}x_1y_1 + a_{12}x_2y_1 + \ldots + a_{21}x_1y_2 + a_{22}x_2y_2 + \ldots + a_{mn}x_ny_m\\
&\quad \geqslant b_1y_1 + b_2y_2 + \ldots .
\end{aligned}
$$

Now, let us consider constraints (6.8). Multiplying the first of these constraints by x_1, the second by x_2, and so on, and adding all the results, we get

$$
\begin{aligned}
&a_{11}x_1y_1 + a_{21}x_1y_2 + \ldots + a_{12}x_2y_1 + a_{22}x_2y_2 + \ldots + a_{mn}x_ny_m\\
&\quad \leqslant c_1x_1 + c_2x_2 + \ldots .
\end{aligned}
$$

The two sums on the left-hand side of these two inequalities are equal to each other. Hence,

$$c_1x_1 + c_2x_2 + \ldots + c_nx_n \geqslant b_1y_1 + b_2y_2 + \ldots + b_my_m,$$

i.e.,

$$Z \geqslant w.$$

In other words, any feasible solution to the primal is greater than or equal to any feasible solution to the dual. Therefore, if the feasible solution to the primal approximates minus infinity, there can be no feasible solution to the dual. Similarly, if the feasible solution to the dual approaches infinity, there can be no feasible solution to the primal. In either case, there is no finite optimum for the primal or the dual. If finite optimal solutions exist for both problems, then it is reasonable to expect that these two solutions will coincide. Absolute proof of this theory will be given in Chapter 10.

If the steps (multiplication and addition) applied to constraint equations (6.7) and (6.8) are applied to Eqs. (6.9) and (6.10) which involve slack variables, we obtain

$$a_{11}x_1y_1 + a_{12}x_2y_1 + \ldots + a_{21}x_1y_2 + a_{22}x_2y_2 + \ldots + a_{mn}x_ny_n$$
$$= b_1y_1 + b_2y_2 + \ldots + x_1'y_1 + x_2'y_2 + \ldots + x_m'y_m,$$
$$a_{11}x_1y_1 + a_{21}x_1y_2 + \ldots + a_{12}x_2y_1 + a_{22}x_2y_2 + \ldots + a_{mn}x_ny_n$$
$$= c_1x_1 + c_2x_2 + \ldots + c_nx_n - x_1y_1' - \ldots - x_ny_n'.$$

Since the sums on the left-hand side of these two equations are equal to each other, we get

$$(b_1y_1 + b_2y_2 + \ldots + b_my_m) + (x_1'y_1 + x_2'y_2 + \ldots + x_m'y_m)$$
$$= (c_1x_1 + c_2x_2 + \ldots + c_nx_n) - (x_1y_1' + x_2y_2' + \ldots + x_ny_n'),$$

i.e.,

$$Z - w = (x_1'y_1 + x_2'y_2 + \ldots + x_m'y_m) + (x_1y_1' + x_2y_2' + \ldots + x_ny_n').$$

When an optimal solution is obtained, let $Z_{\min} = \bar{Z}$ and $w_{\max} = \bar{w}$. Let the corresponding values of the variables be $\bar{x}_1, \bar{x}_2, \ldots, \bar{x}_n, \bar{x}_1', \bar{x}_2', \ldots, \bar{x}_m', \bar{y}_1', \bar{y}_2', \ldots, \bar{y}_m', \bar{y}_1', \bar{y}_2', \ldots, \bar{y}_n'$. According to the duality theorem, $\bar{Z} = \bar{w}$. Hence,

$$Z - \bar{w} = 0 = \bar{x}_1'\bar{y}_1 + \bar{x}_2'\bar{y}_2 + \ldots + \bar{x}_m'\bar{y}_m + \bar{x}_1\bar{y}_1' + \bar{x}_2\bar{y}_2$$
$$+ \ldots + \bar{x}_n\bar{y}_n'. \tag{6.11}$$

Since all the variables are non-negative, each product term in Eq. (6.11) must be zero. Therefore,

$$\bar{x}_j'\bar{y}_j = 0 \qquad (j = 1, 2, \ldots, m).$$

This means that

$$\bar{x}_j' = 0 \quad \text{if } \bar{y}_j > 0,$$
$$\bar{y}_j = 0 \quad \text{if } \bar{x}_j' > 0.$$

Similarly, for

$$\bar{x}_j\bar{y}_j' = 0 \qquad (j = 1, 2, \ldots, n),$$

we have

$$\bar{x}_j = 0 \quad \text{if } \bar{y}_j' > 0,$$
$$\bar{y}_j' = 0 \quad \text{if } \bar{x}_j > 0.$$

This means:

(i) If, in the optimal solution of a primal, a slack variable x_j' has a non-zero value, then the j-th variable of its dual has a zero value in the optimal solution of the dual. A similar statement is true of the dual also.

(ii) If the j-th variable is positive in either system, the j-th constraint of its dual is an equation.

Statements (i) and (ii) are generally known as the *complementary slackness theorem.*

We shall now work out a few problems to demonstrate the primal-dual relationships.

Problem 6.4 Construct the dual of the following linear problem, and solve both the primal and the dual:

$$\text{Minimize } Z = 4x_1 + 2x_2 + 3x_3$$

subject to the constraints

$$2x_1 + 4x_3 \geqslant 5,$$

$$2x_1 + 3x_2 + x_3 \geqslant 4,$$

$$x_1 \geqslant 0,\ x_2 \geqslant 0,\ x_3 \geqslant 0.$$

This is a repetition of Problem 4.2 which has the optimal solution

$$x_1 = 0, \qquad x_2 = 11/12, \qquad x_3 = 5/4, \qquad Z_{\min} = 67/12.$$

The Simplex coefficients in the objective function appear as

$$(-3/2)x_1 - (7/12)x_4 - (2/3)x_5 + Z = 67/12,$$

where x_4 and x_5 are the slack variables.

The dual of this primal is:

$$\text{Maximize } w = 5y_1 + 4y_2$$

subject to the constraints

$$2y_1 + 2y_2 \leqslant 4,$$

$$3y_2 \leqslant 2,$$

$$4y_1 + y_2 \leqslant 3,$$

$$y_1 \geqslant 0,\ y_2 \geqslant 0.$$

The graphical solution to this dual is shown in Fig. 6.1.

The optimal solution is

$$y_1 = 7/12, \qquad y_2 = 2/3, \qquad w_{\max} = 67/12.$$

Now, we can check the validity of the duality theorem according to which

$$Z_{\min} = w_{\max} = 67/12.$$

The values of the basic variables in the optimal solution are $y_1 = 7/12$,

$y_2 = 2/3$; these are numerically equal to the Simplex coefficients of the slack variables in the primal.

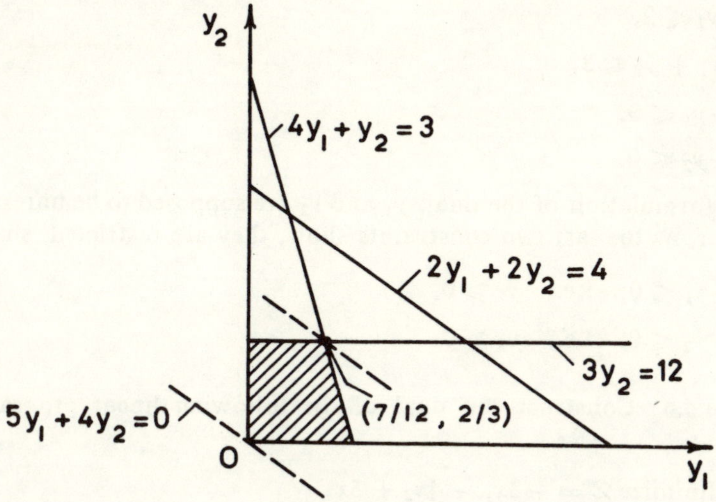

FIGURE 6.1

We can also check Statement (ii) of the complementary slackness theorem. In the primal, the variables x_2 and x_3 are positive. Hence, their corresponding constraint conditions in the dual should be strict equalities, i.e.,

$$3y_2 = 2, \qquad 4y_1 + y_2 = 3.$$

These are true, as is evident in Fig. 6.1. Similarly, since y_1 and y_2 are positive in the dual, their corresponding constraints in the primal must be strict equalities. This is so because

$$4x_3 = 5, \qquad 3x_2 + x_3 = 4.$$

Problem 6.5 Write down the primal of Problem 6.4 with constraint conditions in the form of equalities, and construct its dual:

$$\text{Minimize } Z = 4x_1 + 2x_2 + 3x_3$$

subject to

$$2x_1 + 4x_3 - x_4 = 5,$$

$$2x_1 + 3x_2 + x_3 + x_5 = 4,$$

$$x_1, x_2, x_3, x_4, x_5, \text{ all} \geqslant 0.$$

The dual is:

$$\text{Maximize } w = 5y_1 + 4y_2$$

subject to the constraints

$$2y_1 + 2y_2 \leqslant 4,$$
$$3y_2 \leqslant 2,$$
$$4y_1 + y_2 \leqslant 3,$$
$$-y_1 \leqslant 0,$$
$$-y_2 \leqslant 0.$$

In this formulation of the dual, y_1 and y_2 are supposed to be unrestricted. However, as the last two constraints show, they are restricted, since

$$-y_1 \leqslant 0, \quad \text{i.e.,} \quad y_1 \geqslant 0,$$
$$-y_2 \leqslant 0, \quad \text{i.e.,} \quad y_2 \geqslant 0.$$

Problem 6.6 Construct the dual of the following linear programming problem:

$$\text{Minimize } Z = -3x_1 - 4x_2 + 5x_3$$

subject to

$$3x_1 + x_2 - x_3 = 5,$$
$$2x_1 + 4x_2 - x_3 = 8,$$
$$x_1 \geqslant 0, x_2 \geqslant 0, x_3 \geqslant 0.$$

The solution to this primal is

$$x_1 = 6/5, \qquad x_2 = 7/5, \qquad x_3 = 0, \qquad Z_{\min} = -46/5.$$

The Simplex coefficients are $x_1 = x_2 = 0$, $x_3 = -37/10$.

The dual of this primal is:

$$\text{Maximize } w = 5y_1 + 8y_2$$

subject to the constraints

$$3y_1 + 2y_2 \leqslant -3,$$
$$y_1 + 4y_2 \leqslant -4,$$
$$-y_1 - y_2 \leqslant 5,$$
$$-\infty \leqslant y_1 \leqslant +\infty,$$
$$-\infty \leqslant y_2 \leqslant +\infty.$$

The graphical solution to this dual is shown in Fig. 6.2.

The optimal solution is

$$y_1 = -2/5, \qquad y_2 = -9/10, \qquad w_{\max} = -46/5.$$

As can be seen, the decision variables in the dual are negative, and $Z_{min} = w_{max}$.

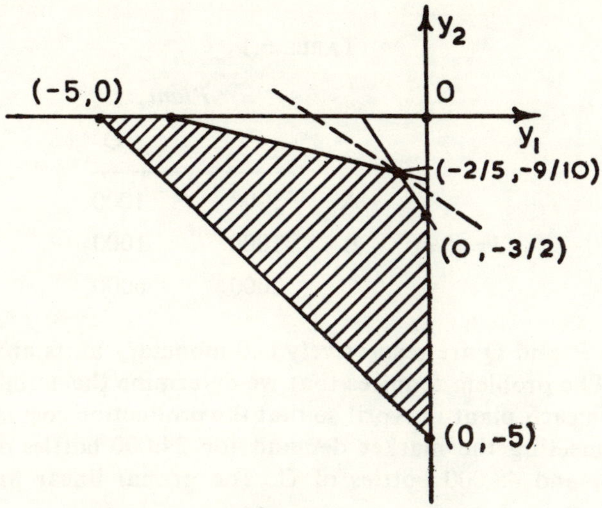

FIGURE 6.2

To show that factor (i) of the duality theorem also holds true, we need the values of the slack variables in the optimal solution of the dual. To obtain these through the Simplex method is quite cumbersome. Instead, we can easily get these values by observing Fig. 6.2. We note that the corner of the solution-polygon giving the optimal solution is the intersection of the first two constraint equations

$$3y_1 + 2y_2 = -3,$$

$$y_1 + 4y_2 = -4.$$

Hence, the slack variables corresponding to these two constraint conditions do not exist. The only constraint condition involving a slack variable is therefore the third one. The equation for the line representing this condition is $y_1 + y_2 = -5$. We know that both y_1 and y_2 are negative in the optimal solution ($y_1 = -2/5$, $y_2 = -9/10$). Hence, the slack variable has the value $-5 + 2/5 + 9/10 = -37/10$. This value is numerically equal to the Simplex coefficient in the primal.

6.4 ECONOMIC INTERPRETATION OF THE DUAL

If the dual has a smaller number of decision variables than the primal, it is more convenient to first solve the dual in order to arrive at the result of the primal. In addition to this advantage, the dual has an interesting interpretation from the economic or cost point of view. To describe this, let us once again consider Problem 1.6 which states that the soft drinks firm has two bottling plants P and Q, each producing three different soft drinks, A, B, and C. The capacities of the two plants, in number of bottles per day, are as given in Table 6.1. The operating costs, per day, of running plants P and Q are respectively 600 monetary units and 400 monetary units. The problem requires that we determine the number of days the firm runs each plant in April so that the production cost is minimized while still meeting the market demand for 24,000 bottles of A, 16,000 bottles of B, and 48,000 bottles of C. The primal linear programming formulation is:

TABLE 6.1

		Plant	
		P	Q
	A	3000	1000
Product	B	1000	1000
	C	2000	6000

$$\text{Minimize } Z = 600x_1 + 400x_2 \tag{6.12}$$

subject to the constraints

$$\begin{aligned} 3000x_1 + 1000x_2 &\geqslant 24{,}000, \\ 1000x_1 + 1000x_2 &\geqslant 16{,}000, \\ 2000x_1 + 6000x_2 &\geqslant 48{,}000, \\ x_1 \geqslant 0,\ x_2 &\geqslant 0, \end{aligned} \tag{6.13}$$

where x_1 and x_2 denote respectively the number of days plants P and Q operate in April. The dual of this primal is:

$$\text{Maximize } w = 24{,}000y_1 + 16{,}000y_2 + 48{,}000y_3 \tag{6.14}$$

subject to the constraints

$$\begin{aligned} 3000y_1 + 1000y_2 + 2000y_3 &\leqslant 600, \\ 1000y_1 + 1000y_2 + 6000y_3 &\leqslant 400, \end{aligned} \tag{6.15}$$

$$y_1, y_2, y_3, \text{ all} \geqslant 0.$$

In constraints (6.15), since the right-hand side denotes monetary units, the left-hand side also should be expressed in monetary units. Let us consider factor $3000y_1$. Here 3000 is the index of the number of bottles per day. Hence, y_1 must be the index of the cost per bottle. Since 3000 is the number of bottles per day of product A, y_1 is the cost per bottle of product A. Similarly, y_2 is the cost per bottle of product B, and y_3 the cost per bottle of product C. These are called the *shadow prices* of products

A, B, and C. They are not the actual market prices, but the true *accounting values* or the *imputed values* of the products.

The objective in the dual is to maximize the total accounting value of the products produced per month. The constraint in the dual dictates that the total accounting value of the daily output of each plant cannot exceed the daily cost of operating the plant. The shadow prices are the values that a company should set on its resources in order to reflect their value to society; and the constraints state that the internal price cannot be set to get more value from a product than that which the company puts into it. In other words, in a situation of equilibrium, the laws of economics for society require no profit.

The solution to the primal problem is (see Problem 1.6) $x_1 = 4$ and $x_2 = 12$, and the minimum production cost, while operating plant P for 4 days and plant Q for 12 days, is 7200 monetary units. The solution to the dual turns out to be $y_1 = 0.1$, $y_2 = 0.3$, and $y_3 = 0$. These are the shadow prices of products A, B, and C, respectively. Since $y_3 = 0$, the accounting value of product C is zero. This means that product C is realized in surplus as a by-product.

To bring out this value aspect, let us consider a blending problem.

Problem 6.7 A large distributing company buys coffee seeds from four different plantations. On these plantations, the seeds are available only in a blend of two types, A and B. The company wants to market a blend of 30% of type A and 70% of type B. The percentage of each type used by each plantation and the selling price per 10 lb of the blends of each plantation are as in Table 6.2. What quantity of coffee seeds should the company buy from each plantation so that the total mixture contains the desired percentages of A and B and at the same time the purchasing cost is at a minimum?

TABLE 6.2

Type	*Plantation*				*Desired %*
	1	2	3	4	
A	40%	20%	60%	80%	30%
B	60%	80%	40%	20%	70%
Selling Price per 10 lb	\$ 3	\$ 2	\$ 1.20	\$ 1.50	Minimum

Let x_1, x_2, x_3, and x_4 be respectively the amounts of coffee seeds from plantations 1, 2, 3, and 4 contained in 1 lb of the distributing company's blend. Then the formulation of the primal problem is:

Minimize $Z = 3x_1 + 2x_2 + 1.2x_3 + 1.5x_4$

subject to

$$0.4x_1 + 0.2x_2 + 0.6x_3 + 0.8x_4 = 0.3,$$

$$0.6x_1 + 0.8x_2 + 0.4x_3 + 0.2x_4 = 0.7,$$

$$x_1, x_2, x_3, x_4, \text{ all} \geqslant 0.$$

The solution is obtained as in Tableau I. With x_2 and x_3 as basic

TABLEAU I

Basis	x_1	x_2	x_3	x_4	Z	$
x_2	1/2	1	0	−1/2		3/4
x_3	1/2	0	1	3/2		1/4
	−7/4	0	0	−7/10		9/5

variables, we have obtained the optimal solution. Hence,

$$x_1 = 0, \quad x_2 = 3/4, \quad x_3 = 1/4, \quad x_4 = 0,$$

$$Z_{\min} = \$\ 9/5 \text{ per } 10 \text{ lb.}$$

The dual of this primal is:

$$\text{Maximize } w = 0.3y_1 + 0.7y_2$$

subject to the constraints

$$0.4y_1 + 0.6y_2 \leqslant 3,$$

$$0.2y_1 + 0.8y_2 \leqslant 2,$$

$$0.6y_1 + 0.4y_2 \leqslant 1.2,$$

$$0.8y_1 + 0.2y_2 \leqslant 1.5.$$

The solution to this dual is

$$y_1 = 2/5, \quad y_2 = 12/5.$$

The interpretation of the dual is as follows: If types A and B were separately available, at what prices should these be sold to the distributing company? According to the shadow prices, type A should cost $ 2/5 per 10 lb, and type B, $ 12/5 per 10 lb. In other words, the true accounting values for the company are $ 2/5 per lb of A and $ 12/5 per lb of B.

6.5 DUAL SIMPLEX METHOD

In Chapter 5, we have shown that a set of basic variables giving a feasible solution can be obtained by using the artificial variables and the infeasibility form. A variant of this, called the *M*-method, has also been discussed. Utilizing the primal-dual relationships for a problem, we have another method for obtaining an initial feasible solution. In many situations, this method, known as the *Dual Simplex method*, enables us to easily identify the initial basis. The method can be described by means of an example.

Problem 6.8

$$\text{Minimize } Z = 4x_1 + 2x_3 \tag{6.16}$$

subject to the constraints

$$\begin{aligned} &x_1 + x_2 - x_3 \geqslant 5, \\ &x_1 - 2x_2 + 4x_3 \geqslant 8, \\ &x_1 \geqslant 0,\ x_2 \geqslant 0,\ x_3 \geqslant 0. \end{aligned} \tag{6.17}$$

Let us consider this problem as the primal problem. Its dual is:

$$\text{Maximize } w = 5y_1 + 8y_2$$

subject to the constraints

$$\begin{aligned} &y_1 + y_2 \leqslant 4, \\ &y_1 - 2y_2 \leqslant 0, \\ &-y_1 + 4y_2 \leqslant 2, \\ &y_1 \geqslant 0,\ y_2 \geqslant 0. \end{aligned}$$

Considering the primal problem, we have Tableau I with slack variables introduced. A starting solution with x_3, x_4, and x_5 as basic variables is not feasible since these will have negative values, -5 and -8. However, the quantities in the last row show that for a minimization problem we have reached the optimality criterion. Therefore, we say that the basis is dual-feasible. In the method proposed, the dual feasibility criterion is maintained, but the primal infeasibility is removed step by step.

TABLEAU I

x_1	x_2	x_3	x_4	x_5	Z	
1	1	−1	−1			5
1	−2	4		−1		8
−4		−2			1	0

The first step in the Dual Simplex algorithm is to remove one of the infeasible basic variables. The *Dual Simplex criterion* (designated as *criterion I*) states that if there are basic variables having negative values, the one that is most negative should be selected. To apply this criterion, the initial tableau is changed to make the basic variables appear with a (+1) coefficient as in Tableau II. According to criterion I, the basic variable to be removed is x_5 since it is more negative than x_4. The next

TABLEAU II

x_1	x_2	x_3	x_4	x_5	Z	
−1	−1	1	1			−5
−1	2	−4		*1*		−8
−4		−2			1	0

question pertains to the new basis to be introduced in place of the one removed. To determine this, we take the ratios of the non-basic variable coefficients (i.e., the coefficients in the row corresponding to the objective function) to the corresponding coefficients in the row of the basic variable that is being removed. In this process, we ignore ratios with zero or positive numbers in the denominator. We then select the variable with the minimum ratio as the coefficient. We shall designate this *Dual Simplex criterion* as *criterion II* and form the *ratios* as in Table 1. The minimum ratio is 1/2 and this corresponds to the variable x_3. Thus, x_3 should

RATIO TABLE 1

non-basic variable	x_1	x_2	x_3
coefficient in last row	−4	0	−2
coefficient in row 2	−1	2	−4
Ratio	4		*1/2*

replace x_5 as the basis. For this we perform, as shown in Tableau III, the regular Simplex operation of getting a (+1) coefficient for x_3 in row 2 and zeros in the other rows.

TABLEAU III

x_1	x_2	x_3	x_4	x_5	Z	
−5/4	−1/2	0	*1*	1/4		−7
1/4	−1/2	1		−1/4		2
−7/2	−1	0		−1/2		4

The corresponding basic solution is

$$Z = 4, \qquad x_4 = -7, \qquad x_3 = 2.$$

Thus, one infeasible solution has been removed. The solution for x_4 is still infeasible since it has a value of -7. To apply criterion II, we construct Table 2 which gives the ratios. The minimum ratio is 2 and

RATIO TABLE 2

non-basic variable	x_1	x_2	x_5
coefficient in last row	$-7/2$	-1	$-1/2$
coefficient in row 1	$-5/4$	$-1/2$	$1/4$
Ratio	$14/5$	2	

this corresponds to the variable x_2. Hence, the variable x_4 leaves the basis, giving place to x_2. Tableau IV shows this change.

TABLEAU IV

x_1	x_2	x_3	x_4	x_5	Z
$5/2$	1	0	-2	$-1/2$	14
$6/4$	0	1	-1	$-1/2$	9
-1	0	0	-2	-1	18

The associated basic solution is

$$Z = 18, \qquad x_2 = 14, \qquad x_3 = 9.$$

The basic variables are all non-negative and all the coefficients in the last row are negative. Hence, this is the optional solution.

In the Dual Simplex method described, it is important to observe the following:

(i) In the initial tableau, the basic variables chosen are infeasible since they are negative.

(ii) All the coefficients appearing in the last row, corresponding to the objective function, possess the correct signs (i.e., negative signs for a minimization problem), thus indicating that the optimality condition has been achieved.

(iii) In the process of removing an infeasible variable and introducing a new non-basic variable in its place, criterion II needs to be applied. For a minimization problem, this criterion requires that we take the ratios of the non-basic coefficients (i.e., the coefficients appearing in the last row corresponding to the objective function) to the coefficients in the row of the basic variable that is being removed; ignore ratios with zero or positive numbers in the denominator; and select the minimum ratio among these, ensuring that the variable corresponding to this minimum ratio enters as the new basis. For a maximization problem, the same procedure is repeated, but the maximum ratio and the corresponding variable are chosen as the new basis. The reason for selecting the minimum ratio in the minimization problem and the maximum ratio in the maximization problem is to stay as close as possible to the optimal value as given in the initial or subsequent tableau.

Problem 6.9

$$\text{Minimize } Z = 3x_1 + x_2 + 2x_3$$

subject to the constraints

$$x_1 + x_2 - 2x_3 \geqslant 1,$$
$$x_1 - 2x_2 + x_3 \geqslant -1,$$

x_1, x_2, x_3 being non-negative.

When introducing slack variables, the constraints become

$$x_1 + x_2 - 2x_3 - x_4 = 1,$$
$$x_1 - 2x_2 + x_3 - x_5 = -1.$$

Let x_4 and x_5 be identified as the basis. Each of these appears with a (+1) coefficient in Tableau I.

TABLEAU I

x_1	x_2	x_3	x_4	x_5	Z
−1	−1	2	*1*		−1
−1	2	−3		1	1
−3	−1	−2			0

The last row indicates that the optimality condition has been reached, but the solution has x_4 as basic-infeasible. To apply criterion II, we form the ratios as in Table 1.

RATIO TABLE 1

non-basic variable	x_1	x_2	x_3
coefficient in last row	−3	−1	−2
coefficient in row 1	−1	−1	2
Ratio	3	*1*	

Choosing the minimum ratio, we get x_2 as the basis. Thus results Tableau II.

TABLEAU II

x_1	x_2	x_3	x_4	x_5	Z
1	1	−2	−1		1
−3	0	1	2	1	−1
−2	0	−4	−1		1

In this process, x_5 has become infeasible. We derive another ratio

RATIO TABLE 2

non-basic variable	x_1	x_3	x_4
coefficient in last row	−2	−4	−1
coefficient in row 2	−3	1	2
Ratio	2/3		

from Table 2. Hence, x_1 enters into the basis. The result is now shown as Tableau III.

TABLEAU III

x_1	x_2	x_3	x_4	x_5	Z
0	1	−5/3	−1/3	1/3	2/3
1	0	−1/3	−2/3	−1/3	1/3
0	0	−14/3	−7/3	−2/3	5/3

The basic solutions are now feasible and the optimality condition has also been reached. Hence, the solution is

$$x_1 = 1/3, \quad x_2 = 2/3, \quad Z_{\min} = 5/3.$$

EXERCISES

1. Solve the following primal graphically; also, write down its dual and solve this too graphically:

Maximize $Z = x_1 + 5x_2$

subject to the constraints

$$5x_1 + 6x_2 \leqslant 30,$$

$$3x_1 + 2x_2 \leqslant 12,$$

$$x_1, x_2 \geqslant 0.$$

[*Ans.* $Z_{max} = 25, x_1 = 0, x_2 = 5$.]

2. Write down the dual of Exercise 1 (Chapter 4). Solve it and check the validity of the duality theorem. Check also the complementary slackness theorem.
[*Ans.* *Primal*: $x_1 = 18, x_2 = x_3 = x_4 = 0$; slack: $x_1' = 14, x_2' = 0, x_3' = 54$. *Dual*: $y_1 = 0, y_2 = 3, y_3 = 0$; slack: $y_1' = 0, y_2' = 2, y_3' = 9, y_4' = 11, w_{min} = 108$.]

3. Interpret the dual in Exercise 2 in economic terms. What will the basic variables in the dual now represent?
[*Ans.* The variables in the dual represent the cost per hour of operating machines A and B, and the optical grinder G. As the solution to the dual indicates, the operating cost is: A: nil; B: Rs 3 per hour; G: nil.]

4. Write down the dual of Exercise 6 (Chapter 4) and solve it by means of the duality and the complementary slackness theorems. Make use of the solution to the primal problem.
[*Ans.* *Primal*

$$\text{Maximize } Z = 6x_1 + 4x_2 + 7x_3 + 5x_4$$

subject to the constraints

$$6x_1 + 5x_2 + 3x_3 + 2x_4 \leqslant 100,$$

$$3x_1 + 4x_2 + 9x_3 + 2x_4 \leqslant 75,$$

$$x_1 + 2x_2 + x_3 + 2x_4 \leqslant 20,$$

$$x_1, x_2, x_3, x_4, \text{ all} \geqslant 0.$$

Dual

$$\text{Minimize } w = 100y_1 + 75y_2 + 20y_3$$

subject to the constraints

$$6y_1 + 3y_2 + y_3 \geqslant 6,$$

$$5y_1 + 4y_2 + 2y_3 \geqslant 4,$$

$$3y_1 + 9y_2 + y_3 \geqslant 7,$$

$$2y_1 + 2y_2 + 2y_3 \geqslant 5,$$

$$y_1, y_2, y_3, \text{ all} \geqslant 0.$$

The solution to the primal is $x_1 = 15$, $x_2 = 0$, $x_3 = 10/3$, $x_4 = 0$. Hence, the slack variables in the primal are $x_1' = 0$, $x_2' = 0$, $x_3' = 5/3$. Consequently, in the dual, $y_1 \neq 0$, $y_2 \neq 0$, $y_3 \neq 0$. Further, since in the primal $x_1 \neq 0$, $x_2 = 0$, $x_3 \neq 0$, $x_4 = 0$, the slack variables in the dual are $y_1' = 0$, $y_2' \neq 0$, $y_3' = 0$, $y_4' \neq 0$. This means that the first constraint in the dual is an equality, the second constraint an inequality, the third constraint an equality, and the fourth constraint an inequality. This, coupled with the result that $y_3 = 0$, yields from the dual

$$6y_1 + 3y_2 = 6,$$

$$3y_1 + 9y_2 = 7.$$

Solving these, we get $y_1 = 11/15$, $y_2 = 8/15$, $w_{\min} = 113\frac{1}{3}$.]

5. Give the economic interpretation of the dual as stated in Exercise 4.
[*Ans.* In the dual, y_1 = cost of steel per kg, y_2 = cost of brass per kg, y_3 = cost of labour per week.]

6. Solve the following primal; also write down its dual and solve it:
Primal

$$\text{Maximize } Z = 10x_1 - x_2 - 9x_3 - 8x_4$$

subject to

$$2x_1 - x_2 - 3x_3 - x_4 + 2 = 0,$$

$$5x_1 - 2x_2 - 3x_4 + 5 = 0,$$

$$-7x_1 + 4x_2 - x_3 - 4x_4 + 1 \geqslant 0,$$

$$-3x_1 - 2x_2 - 5x_3 - 6x_4 + 10 \geqslant 0,$$

$$x_1, x_2, x_3, x_4, \text{ all } \geqslant 0.$$

[*Ans.* *Primal*: $Z_{\max} = -62/7$ with $x_1 = 1/7$, $x_2 = 8/7$, $x_3 = 0$, $x_4 = 8/7$. *Dual*: $w_{\min} = -62/7$ with $u_1 = 111/7$, $u_2 = -57/7$, $u_3 = 1/7$, $u_4 = 0$. *Note*: In the dual, u_1 and u_2 are not bound by the non-negativity constraint since the first two constraints in the primal are equalities.]

7. Write down the dual of Exercise 7, Chapter 4, and solve it.
[*Ans.* $w_{\min} = 18$ with $u_1 = -109/3$, $u_2 = 15$, $u_3 = 92/3$.]

8. Use the Dual Simplex method to

$$\text{minimize } Z = 2y_3 + y_4 + 2y_5$$

subject to

$$y_3 + y_4 - 2y_5 - y_1 = 1,$$

$$y_3 - 2y_4 + y_5 - y_2 = -1,$$

all y's being non-negative.
[*Ans.* $Z = 5/3$, $y_3 = 1/3$, $y_4 = 2/3$, $y_5 = 0$.]

9. Minimize $Z = 4x_1 + 2x_2 + 3x_3$
subject to the constraints

$$x_1 - 2x_2 + x_3 \leqslant 5,$$

$$2x_1 + 3x_2 - x_3 = 2,$$

$$x_1 - 5x_2 + 6x_3 \geqslant 3,$$

$$x_1, x_2, x_3 \geqslant 0.$$

[*Ans.* $Z = \frac{72}{13}$, $x_1 = \frac{15}{13}$, $x_2 = 0$, $x_3 = \frac{4}{13}$.]

7

Applications

7.1 INTRODUCTION

In this chapter, we shall consider the applications of linear programming techniques to four specific situations and their variants. These are:

(1) Transportation Model
(2) Linear Production Model
(3) Critical Path Scheduling
(4) Allocation Problem

Though we can apply the Simplex or the Dual Simplex procedure in all these situations, yet their particular nature makes possible a more direct method or variations of the Simplex procedure. We shall deal with these variations in each case.

7.2 TRANSPORTATION MODEL

7.2.1 Classical Problem

Let us consider the following example which deals with the production of goods and their shipment from one place to another.

Problem 7.1 A company has three manufacturing plants situated at A, B, and C, with production capacities of respectively 2000, 6000, and 7000 units per week. These units are to be shipped to four distributing centres, D, E, F, and G, with absorption capacities of respectively 3000, 3000, 4000, and 5000 units per week. The company would like to distribute the manufactured items in such a way that the total shipping cost from the plants to the distributing centres is a minimum. The various shipping costs, per unit product shipped, from the plants to the centres are given in Table 7.1.

Let the number of units shipped from plant A to centres D, E, F, and G be x_{11}, x_{12}, x_{13}, and x_{14}, respectively. Similarly, from plants B and C to the four centres. The number of units shipped from plant i to the distribution centre j will, in general, be represented by x_{ij}. The cost of shipping one unit of the product from plant i to centre j will be represented by c_{ij}. If the production capacities of the plants are represented by $a_i (i = 1, 2, 3)$, and the distributing capacities of the centres by b_j $(j = 1, 2, 3, 4)$, the

TABLE 7.1

	To	D	E	F	G
	A	13	11	15	20
From	B	17	14	12	13
	C	18	18	15	12

transportation problem becomes:

$$\text{Minimize } Z = c_{11}x_{11} + c_{12}x_{12} + \ldots + c_{34}x_{34} \tag{7.1}$$

subject to

$$x_{11} + x_{12} + x_{13} + x_{14} = a_1, \tag{7.2a}$$

$$x_{21} + x_{22} + x_{23} + x_{24} = a_2, \tag{7.2b}$$

$$x_{31} + x_{32} + x_{33} + x_{34} = a_3, \tag{7.2c}$$

$$x_{11} + x_{21} + x_{31} = b_1, \tag{7.2d}$$

$$x_{12} + x_{22} + x_{32} = b_2, \tag{7.2e}$$

$$x_{13} + x_{23} + x_{33} = b_3, \tag{7.2f}$$

$$x_{14} + x_{24} + x_{34} = b_4. \tag{7.2g}$$

Using the abbreviated form, we can state the problem as:

$$\text{Minimize } Z = \sum_{i=1}^{3} \sum_{j=1}^{4} c_{ij}x_{ij}$$

subject to

$$\sum_{j=1}^{4} x_{ij} = a_i,$$

$$\sum_{i=1}^{3} x_{ij} = b_j,$$

$$x_{ij} \geqslant 0.$$

This is a problem in linear programming possessing certain peculiarities. All the coefficients in the constraint equations are either unity or zero and the pattern of variables, occurring in any single equation, is a very special one. The total production capacity of all the three plants is equal to the total distribution capacity of all the four centres, i.e.,

$$a_1 + a_2 + a_3 = b_1 + b_2 + b_3 + b_4.$$

Equivalently,

$$\text{Eqs. } [(7.2a) + (7.2b) + (7.2c)] - [(7.2d) + (7.2e) + (7.2f)] = \text{Eq. } (7.2g).$$

Hence, among the seven constraint equations, only six are independent; consequently, a basic-feasible solution will consist of not more than six positive values. To find such a feasible solution, we adopt what is known as the *north-west corner rule.*

7.2.2 North-West Corner Rule and Stepping-Stone Method

Let us consider Table 7.2 which we shall fill in. The marginal values (also known as *rim requirements*) give the plant capacities and the demand requirements of the various distribution centres.

TABLE 7.2

		Distribution Centre Demand			
		D	E	F	G
		3000	3000	4000	5000
Plant Capacity	A 2000				
	B 6000				
	C 7000				

We first fill in the north-west corner, i.e., the cell corresponding to x_{11}. If the capacity of plant A exceeds the requirement of centre D, then we fill in the complete quota for centre D and distribute the remaining units of plant A to the other centres. The attempt in distributing to other centres is to meet their required quotas to the maximum extent possible. On the other hand, if the production capacity of plant A is less than the distribution capacity of centre D, as in Table 7.2, we assign the complete capacity of A to centre D, then try to meet the remaining requirement of centre D first from plant B and then from plant C. In the present example, centre D takes all the 2000 units from A, thereby exhausting the capacity of A. Hence, $x_{12} = x_{13} = x_{14} = 0$. The remaining requirement of centre D (i.e., 1000 units) is obtained from plant B. This leaves 5000 units at B to be distributed to centres E, F, and G. Since we attempt to meet the requirement to the maximum extent, centre E gets 3000 units (which is its

TABLE 7.3

	D	E	F	G
	3000	3000	4000	5000
A 2000	2000			
B 6000	1000	3000	2000	
C 7000			2000	5000

complete quota), and the balance of 2000 units goes to centre F. However, F needs a total of 4000 units. Hence, it gets its balance requirement of 2000 units from plant C. The 5000 units left with C are sent to centre G. Since the total production capacity is equal to the total distribution capacity, the quantities as distributed balance one another. The result is shown in Table 7.3.

This provides us with a basic-feasible solution with six positive values. The total shipping cost of this scheme is

$$\begin{aligned} Z &= (13 \times 2000) + (17 \times 1000) + (14 \times 3000) + (12 \times 2000) \\ &\quad + (15 \times 2000) + (12 \times 5000) \\ &= 199{,}000. \end{aligned}$$

To investigate whether this is the best possible solution, we shall use one of the empty cells as a basic variable to check whether the resulting shipping cost increases or decreases. Let us take x_{12}. If we enter one unit in this cell, we must take away one unit from x_{11}. To meet the requirement of D, we must add one unit to x_{21} which comes from x_{22}. This moving from one cell to another is called the *stepping-stone method.* The total cost, as a consequence, is changed by

$$c_{12} - c_{22} + c_{21} - c_{11} = 1. \tag{7.3}$$

Hence, this is not an economical proposition.

Let us try filling in x_{31}. This changes the cost by

$$c_{31} - c_{21} + c_{23} - c_{33} = -2, \tag{7.4}$$

i.e., the cost of shipping would decrease. In this way, we can examine each empty cell to see whether the scheme can be improved. We notice that by means of this stepping-stone method we complete a circuit by horizontal and vertical steps, leading from an empty cell to three occupied cells (because we are considering only basic-feasible solutions). Adopting the solution suggested by Eq. (7.4), we try to fill in cell 3-1 (i.e., row 3, column 1) in place of the original basis occurring in cell 2-1 (i.e., row 2, column 1). The scheme appears as in Table 7.4. The shipping cost now is 197,000 monetary units.

TABLE 7.4

		D	E	F	G
		3000	3000	4000	5000
A	2000	2000			
B	6000		3000	3000	
C	7000	1000		1000	5000

Let another iteration involving the cells 1-1, 1-2, 2-2, 2-3, 3-3, and 3-1 be made. The shipping cost is changed by

$$-c_{11} + c_{12} - c_{22} + c_{23} - c_{33} + c_{31} = -1$$

and therefore improves the scheme. The resulting scheme is as in Table 7.5. The total shipping cost now is 196,000 monetary units.

TABLE 7.5

		D	E	F	G
		3000	3000	4000	5000
A	2000	1000	1000		
B	6000		2000	4000	
C	7000	2000			5000

Instead of starting with the north-west corner cell, we can adopt the slightly improved approach which we proceed to describe.

7.2.3 Matrix Minimum Method

First, we take the cell with the lowest shipping cost. This is cell 1-2 which has a cost of 11. We fill in this cell first with as many manufactured units as possible, without exceeding its requirement, from the corresponding plant which in this case has 2000 units. The capacity of plant A is thus exhausted and the requirement of centre E is now 1000. This distribution reduces the pattern to that shown in Table 7.6.

TABLE 7.6

		D	E	F	G
		3000	1000	4000	5000
B	6000				
C	7000				

From Table 7.6, we select the cell with the lowest shipping cost and proceed as before. There are now two cells 2-3 and 3-4, both with a cost coefficient of 12. We arbitrarily select cell 3-4 and meet its need from plant C. Proceeding in this manner, we obtain the final scheme as shown in Table 7.7.

TABLE 7.7

		D	E	F	G
		3000	3000	4000	5000
A	2000		2000		
B	6000	1000	1000	4000	
C	7000	2000			5000

The total shipping cost for this scheme is 197,000 monetary units which is less than the solution of the previous scheme, namely, 199,000 monetary units. From this basic scheme, we can proceed with the stepping-stone method. However, a short-cut to this method is possible, and we shall now discuss it.

7.2.4 Fictitious Cost Method

On the margins of the table, we enter numbers: u_i corresponding to the rows $i = 1, 2, 3$; and v_j corresponding to the columns $j = 1, 2, 3, 4$. These numbers satisfy for all *occupied cells* the condition

$$u_i + v_j = c_{ij}. \tag{7.5}$$

These numbers are called *fictitious costs* or *shadow costs*, and they are filled in as follows. (In general, these fictitious costs need not be non-negative.)

Let us choose, arbitrarily, $u_1 = 0$. This automatically fixes the other numbers, because of Eq. (7.5). This is possible only when the number of independent equations in the transportation problem is equal to the number of occupied cells. The values of c_{ij} appearing in the occupied cells and the corresponding values of u_i's and v_j's are as in Table 7.8.

TABLE 7.8

	D	E	F	G	u_i
		11			0
	17	14	12		3
	18			12	4
v_j	14	11	9	8	

Now, let us start with the basic solution as given in Table 7.7 to see whether the introduction of x_{11} into the basis improves the situation. The introduction of x_{11} changes the cost of shipping by

$$c_{11} - c_{12} + c_{22} - c_{21}.$$

If we substitute for c_{12}, c_{22}, and c_{21} in terms of u_i and v_j, and since Eq. (7.5) holds for these costs (as they represent the occupied cells), the change in the cost of shipping becomes

$$c_{11} - (u_1 + v_2) + (u_2 + v_2) - (u_2 + v_1) = c_{11} - (u_1 + v_1). \tag{7.6}$$

Now, if Eq. (7.6) is negative, the new scheme improves the situation. Otherwise, it does not. The reason for using the name *fictitious cost* is now obvious. Equation (7.5), namely, $u_i + v_j = c_{ij}$, is true only for an occupied cell. For an unoccupied cell, this relationship is not true in general and it will therefore be called the fictitious cost. In the present case, $(u_1 + v_1) = 14$ is the fictitious cost for cell 1-1, because the true cost for this cell is 13. According to Eq. (7.5), if the fictitious cost of a cell is greater than the true cost, then introducing that cell into the basis improves the situation. As $u_1 + v_1$ (=14) is greater than c_{11} (=13), it is worthwhile to consider filling in cell 1-1. In this way, we can fill in the *unoccupied cells* with the appropriate fictitious costs and compare them with their true costs. The fictitious costs appear as in Table 7.9.

TABLE 7.9

	D	E	F	G	u_i
	14		9	8	0
				11	3
		15	13		4
v_j	14	11	9	8	

A comparison of the fictitious costs given in Table 7.9 with the true costs given in Table 7.1 shows that it is worthwhile considering x_{11}. Filling in this cell by alternatively adding to and subtracting from x_{11}, x_{21}, x_{22}, and x_{12}, we get the scheme in Table 7.10. This gives a total shipping cost of 196,000 monetary units.

TABLE 7.10

		D	E	F	G
		3000	3000	4000	5000
A	2000	1000	1000		
B	6000		2000	4000	
C	7000	2000			5000

To see whether this is the optimal solution, we proceed to form the fictitious costs. Starting with $u_1 = 0$, we fill in the table for other u_i's and

v_j's, satisfying Eq. (7.5) for all the *newly occupied cells*. This gives the pattern in Table 7.11. The fictitious costs for the *unoccupied cells* are as

TABLE 7.11

	D	E	F	G	u_i
	13	11			0
		14	12		3
	18			12	5
v_j	13	11	9	7	

shown in Table 7.12. The fictitious costs for all the unoccupied cells are

TABLE 7.12

	D	E	F	G	u_i
			9	7	0
	16			10	3
		16	14		5
v_j	13	11	9	7	

smaller than the true costs; hence, the scheme given in Table 7.10 is the optimal solution.

Problem 7.2 Solve the transportation problem given in Table 7.13. The marginal values give the capacities at the sources and the requirements at the various destinations. The other values are the costs of transportation per unit between stations.

TABLE 7.13

			Distribution Centre Demand			
			D	E	F	G
			4	8	3	6
Plant Capacity	A	8	5	3	1	9
	B	4	5	7	3	2
	C	9	3	2	1	8

To obtain the initial basic solution, we adopt the matrix minimum procedure. If we start with cell 13, the solution appears as in Table 7.14.

TABLE 7.14

		D	E	F	G
		4	8	3	6
A	8	3		3	2
B	4				4
C	9	1	8		

Next, we form the shadow costs based on the occupied cells as in Table 7.15.

TABLE 7.15

	D	E	F	G	u_i
A	5		1	9	0
B				2	−7
C	3	2			−2
v_j	5	4	1	9	

The shadow costs for the unoccupied cells in Table 7.15 are as in

TABLE 7.16

	D	E	F	G	u_i
A		4			0
B	−2	−3	−6		−7
C		2	−1	7	−2
v_j	5	4	1	9	

Table 7.16. Comparing these shadow costs with the true costs, we find that only cell 12 competes. For this,

$$c_{12} - (u_1 + v_2) = 3 - 4 = -1.$$

Since this is negative, it is worth filling in cell 12. Considering cells 11, 12, 32, and 31, we obtain the next basic solution as shown in Table 7.17.

TABLE 7.17

	4	8	3	6
8		3	3	2
4				4
9	4	5		

The shadow costs for the unoccupied cells in Table 7.17 are given in Table 7.18.

TABLE 7.18

	D	E	F	G	u_i
A		3	1	9	0
B				2	−7
C	3	2			−1
v_j	4	3	1	9	

The shadow costs for the unoccupied cells in Table 7.18 are then obtained as in Table 7.19. Comparing these shadow costs with the true

TABLE 7.19

	D	E	F	G	u_i
A	4				0
B	−3	−4	−6		−7
C			0	8	−1
v_j	4	3	1	9	

costs, we observe that in cell 34 the true cost is equal to the shadow cost. This means that the solution given in Table 7.19 is optimal with a total cost of 60 monetary units, and we can fill in cell 34 without changing the optimal cost. To fill in cell 34, we have to consider cells 12, 14, 34, and 32. The maximum number of units that can fill in cell 34 is 2; the minimum is zero. If x number of units are put in cell 34 (where $0 \leqslant x \leqslant 2$), the optimal solution to the problem is as shown in Table 7.20, where $0 \leqslant x \leqslant 2$.

TABLE 7.20

		D	E	F	G
		4	8	3	6
A	8		$(3+x)$	3	$(2-x)$
B	4				4
C	9	4	$(5-x)$		x

7.2.5 General Transportation Model

We have assumed in the classical problem that the total capacity of the manufacturing plants equals the sum of the capacities of the distribution centres. This may not happen in general. First, the manufacturing plants may produce less than their capacity levels and, second, whatever is produced may not be taken by the distribution centres. To include these possibilities, let a_i represent the maximum production capacity of plant i, and b_j, the maximum number of units that can be accepted by distribution centre j. The actual number of units taken by centre j may be less than this. Let x_{ij} represent the number of manufactured units supplied by plant i to centre j. As before, the cost of shipping is c_{ij}, representing the cost of shipping one unit of the manufactured item from plant i to centre j. The algebraic formulation of the general transportation problem is then:

$$\text{Minimize } Z = \sum_i \sum_j c_{ij}x_{ij} \tag{7.7}$$

subject to the constraints

$$\sum_j x_{ij} \leqslant a_i, \tag{7.8}$$

$$\sum_i x_{ij} \leqslant b_j. \tag{7.9}$$

We can change the inequalities into equalities as

$$\begin{aligned} \sum_j x_{ij} + x_i &= a_i, \\ \sum_i x_{ij} + y_j &= b_j. \end{aligned} \tag{7.10}$$

In addition, we can write two more equations without further restricting the variables. These are

$$\begin{aligned} \sum_i x_i + z &= \sum_i a_i, \\ \sum_j y_j + z &= \sum_j b_j, \end{aligned} \tag{7.11}$$

where z is the sum of all the x_{ij}'s. The transportation scheme now appears as in Table 7.21.

TABLE 7.21

	b_1	b_2	...	b_j	$\sum_i x_i$
a_1	x_{11}	x_{12}	...	x_{1j}	x_1
a_2	x_{21}	x_{22}	...	x_{2j}	x_2
⋮	⋮	⋮		⋮	⋮
a_i	x_{i1}	x_{i2}	...	x_{ij}	x_i
$\sum_j b_j$	y_1	y_2	...	y_j	z

As an example, let us consider Problem 7.3.

Problem 7.3 An international company manufacturing sewing-machines has two plants, one in India, with a capacity of 200 units per week, and one in Hong Kong, with a capacity of 100 units per week. The company markets its machines to four countries, namely, Hong Kong, India, Egypt, and Ceylon. These countries have a maximum demand of 75, 100, 100, and 30 units, respectively, per week. Due to differences in cost of materials, customs duty, transportation charges, etc., the profit per unit differs according to the country where it is produced and the country where it is sold. These profits are as in Table 7.22. The problem is to plan the production programme so as to *maximize* the profit. The company may have its production capacity at both plants partly or wholly unused.

TABLE 7.22

	Hong Kong	*India*	*Egypt*	*Ceylon*
Hong Kong	80	80	120	100
India	40	60	90	75

Let the unused capacity of the plant in Hong Kong be x_1 and of the one in India, x_2. Let x_{ij} be the number of units supplied by plant i ($i = 1, 2$) to country j ($j = 1, 2, 3, 4$). Let the unsatisfied demand of country j be y_j. The transportation scheme appears as in Table 7.23.

TABLE 7.23

		Hong Kong 75	*India* 100	*Egypt* 100	*Ceylon* 30	S Σa_i
Hong Kong	100	x_{11}	x_{12}	x_{13}	x_{14}	x_1
India	200	x_{21}	x_{22}	x_{23}	x_{24}	x_2
P	Σb_j	y_1	y_2	y_3	y_4	Z*

* $Z = \sum_{i=1}^{2} \sum_{j=1}^{4} x_{ij}$.

This is similar to Table 7.21.

The interpretation of Table 7.23 in terms of the classical transportation model is as follows: when a plant's capacity is not fully utilized, we assume that the plant still works up to its maximum capacity and ships whatever is not sold to a fictitious destination S without any profit. Similarly, when a country's demand is not fully satisfied, it still gets its maximum demand, obtaining its unsatisfied demand from a fictitious plant P. The profit per unit appears as in Table 7.24.

TABLE 7.24

	Hong Kong	*India*	*Egypt*	*Ceylon*	S
Hong Kong	80	80	120	100	0
India	40	60	90	75	0
P	0	0	0	0	0

We shall solve this problem by using the matrix minimum method. In this case, as we are dealing with a maximization problem, we select from the profit table the cell with the largest entry. This is cell 1-3, with a profit index of 120. The Egyptian market can take a maximum of 100 units which can be completely supplied by the Hong Kong plant. This gives the reduced demand-and-supply as in Table 7.25.

TABLE 7.25

		Hong Kong	*India*	*Ceylon*	S
		75	100	30	Σa_i
India	200				
P	Σb_j				

The cell in Table 7.25 with the maximum profit index is the one corresponding to India-Ceylon. Since Ceylon needs a maximum of 30 units only, this can be completely supplied by India. Proceeding in this manner, we get the scheme in Table 7.26.

TABLE 7.26

		Hong Kong	*India*	*Egypt*	*Ceylon*	S
		75	100	100	30	Σa_i
Hong Kong	100			100		
India	200	70	100		30	
P	Σb_j	5				300

Table 7.26 provides also the optimal solution. The demand of the Hong Kong market is left unsatisfied by 5 units. The profit due to this scheme is therefore

$$Z = (100 \times 90) + (70 \times 40) + (100 \times 60) + (30 \times 75)$$
$$= 20{,}050 \text{ monetary units per week.}$$

7.3 TRANS-SHIPMENT MODEL

In the transportation problems discussed so far, we have assumed that there are a certain number of manufacturing plants or supply centres and a given number of warehouses or demand points. In Fig. 7.1, the supply centres are S_1, S_2, and S_3 and the demand points are D_1, D_2, and D_3. The capacities of the supply centres and the requirements of the demand points are indicated by the numbers appearing alongside these points. We note from these figures that the demand and supply need not be equal in a general transportation model. The cost of shipping an item from supply centre S_i to demand point D_j is given by c_{ij}.

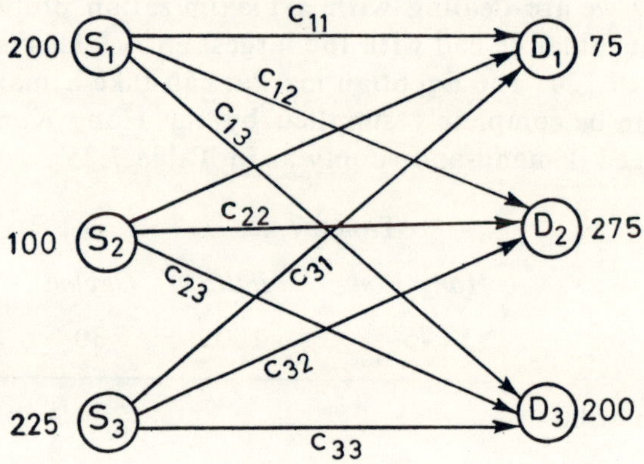

FIGURE 7.1

Two points require attention in Fig. 7.1:

(i) The flow of commodity is from the supply points to the demand points; there is no flow between the demand points themselves.

(ii) Because of (i), the cost of shipping is from S_i to D_j.

In practical situations, the flow pattern for the commodities and the cost structure are quite involved, as is evident from the example that follows.

A manufacturing concern has its production units at three major plants, P_1, P_2, and P_3, and ships its products to several warehouses and retail distributors. The retail distributors (D_1, D_2, D_3, and D_4) can get their requirements either directly from the plants or from the warehouses (W_1, W_2, and W_3). A warehouse can get its requirements either from the plants or from other warehouses.

The foregoing problem may be depicted as in Fig. 7.2. It is an extension of the conventional transportation problem. The requirement D_2, for example, is met partially from plant P_3 and partially from warehouse W_2. The cost structure is still unidirectional in the sense that the cost is from

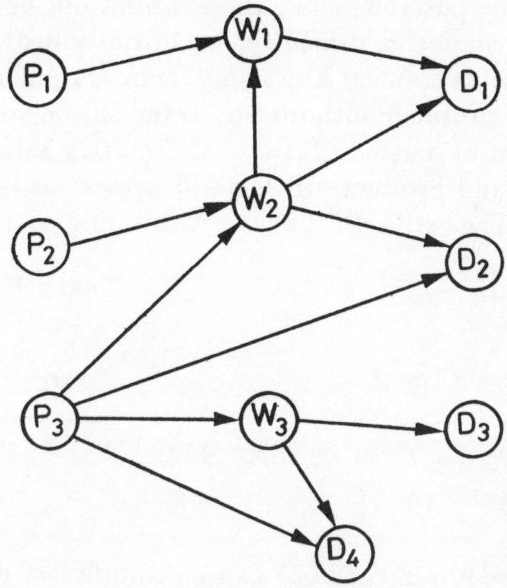

FIGURE 7.2

P_1 to W_1, W_2 to D_1, and so on. If the product can flow from W_2 to W_1 as well as from W_1 to W_2, then the cost structure for the flow between W_1 and W_2 can very well depend on the direction of flow. Problems of this type may be classified as *trans-shipment problems*.

An example of a general trans-shipment problem follows.

Problem 7.4 A manufacturer owns two plants, A and B, with annual production capacities of 900 units and 600 units, respectively. These units are shipped to three ports, C, D, and E, whose requirements are respectively 700, 400, and 300 units. The costs of transport per unit are as indicated in Table 7.27. How should the manufacturer distribute the product shipment?

TABLE 7.27

	To	C	D	E	A	B
	A	9	10	7	0	2
	B	7	5	10	2	0
From	C	0	3	4	9	7
	D	3	0	2	10	5
	E	4	2	0	7	10

In this problem, the cost of shipment between any two stations is the same in both directions. In solving problems involving possibilities of trans-shipment, we cannot adopt the same techniques as used in an ordinary transportation problem. For example, if the requirements of a port, say, D, are fulfilled exactly, then no trans-shipment from D to the other ports would be possible, since there would not be any surplus units available. To overcome this difficulty, we initially provide extra units to all station points so as to meet any likely trans-shipment. Before this is done, we solve the problem without any trans-shipment.

To find the initial basic solution, the matrix minimum method is adopted. The cost and production demand appear as in Tables 7.28 and 7.29, respectively. The cell with the lowest shipping cost is 2-2. This cell is filled in first. The requirement of D can be completely met from plant B. In the resulting pattern, cells 1-2 and 2-1 have the next lower cost. Choosing cell 2-1, we get the final pattern as in Tables 7.30 and 7.31. The total

TABLE 7.28

	C	D	E
A	9	10	7
B	7	5	10

TABLE 7.29

	C	D	E
	700	400	300
A 900			
B 600			

TABLE 7.30

		C	E
A	900	9	7
B	200	7	10

TABLE 7.31

	C	D	E
	700	400	300
A 900	500		300
B 600	200	400	

shipping cost for this solution is

$$4500 + 2100 + 1400 + 2000 = 10{,}000 \text{ monetary units.}$$

To find the optimal solution without trans-shipment, we adopt the fictitious cost (or shadow cost) method. The first table we get is Table 7.32.

TABLE 7.32

	D	E	F	u_i
	9		7	0
	7	5		−2
v_i	9	7	7	

The fictitious costs for the unoccupied cells in Table 7.32 are as in Table 7.33. Comparing the fictitious costs in Table 7.33 with the true

TABLE 7.33

	D	E	F	u_i
		7		0
			5	−2
v_i	9	7	7	

costs in Table 7.28, we observe that

$$c_{12}-(u_1+v_2)=10-7=3, \qquad c_{23}-(u_2+v_3)=10-(7-2)=5.$$

Since these are positive, the transportation solution obtained without trans-shipment is the optimal.

Next, to solve the problem with trans-shipment, we add a large number, say, 2000, to the existing plant capacities as well as to the requirements of the three ports. The ports also start with an initial supply of 2000. This addition provides possibilities for trans-shipment, as is clear from Table 7.34. We observe that the addition of 2000 units to all the ports

TABLE 7.34

		C	D	E	A	B	P	
		2700	2400	2300	2000	2000	100	u_i
A	2900	500		300	2000		100	0
B	2600	200	400			2000		−2
C	2000	2000						−9
D	2000		2000					−7
E	2000			2000				−7
	v_i	9	7	7	0	2	0	

does not alter the original optimal solution. The capacity of plant A has become 2900 units; of these, 500 units are shipped to port C, 300 units to E, 2000 units to A (i.e., 2000 units remain at plant A), and 100 units to a fictitious port P. The figures in the other cells can be similarly interpreted.

To check whether the pattern in Table 7.34 is the optimal one, we formulate the fictitious costs for the unfilled cells, as in Table 7.35. For these cells, the difference between the actual cost (given in Table 7.27) and the fictitious cost can be determined. If the solution given in

TABLE 7.35

	C	D	E	A	B	P	u_i
		7			2		0
			5	−2		−2	−2
		−2	−2	−9	−7	−9	−9
	2		0	−7	−5	−7	−7
	2	0		−7	−5	−7	−7
v_j	9	7	7	0	2	0	

Table 7.34 is not optimal, then there would be cells in Table 7.35 for which $[(c_{ij} - (u_i + v_j)]$ would be negative, and filling in these cells would give a better solution. If there is more than one cell satisfying this requirement, then we would fill in that cell which would have the most negative value. In the present case, there is no cell in Table 7.35 whose fictitious cost is greater than the corresponding true value. Hence, the solution given in Table 7.34 is an optimum one. However, cell 15 (corresponding to AB) has a fictitious cost (=2) which is equal to the true cost (=2). This means that filling in this cell appropriately also yields an optimal solution with no change in the transportation cost. But filling in this cell means trans-shipment. First, we select cells 11, 15, 25, and 21. If x is the number of units introduced into cell 15, the units would be as shown in Table 7.36, where $0 \leqslant x \leqslant 500$.

TABLE 7.36

		C	D	E	A	B	P	
		2700	2400	2300	2000	2000	100	u_i
A	2900	$(500-x)$		300	2000	x	100	0
B	2600	$(200+x)$	400			$(2000-x)$		−2
C	2000	2000						−9
D	2000		2000					−7
E	2000			2000				−7
	v_j	9	7	7	0	2	0	

Table 7.36 contains a surplus of 2000 units which were artificially introduced to provide for trans-shipment. To extract from this the optimal transportation schedule, we omit the entries in diagonal cells (i.e., AA, BB, CC, DD, EE) and obtain Table 7.37, where $0 \leqslant x \leqslant 500$. According to the optimal solution obtained, there is shipment from A to

TABLE 7.37

		C	D	E	A	B
		700	400	300		
A	900	$(500-x)$		300		x
B	600	$(200+x)$	400			

B, and consequently B is able to supply more than what it actually produces. Alternatively, we can say that port C receives a part of its shipment from A via B.

The total cost of trans-shipment is given by

$$\begin{aligned} Z &= (500 - x)9 + 300(7) + 2x + 400(5) + (200 + x)7 \\ &= 4500 - 9x + 2100 + 2x + 2000 + 1400 + 7x \\ &= 10{,}000 \text{ monetary units.} \end{aligned}$$

This is independent of x as it should be, since x can vary from zero value (i.e., no trans-shipment) to a maximum of 500 (i.e., all shipment from A to C is via B only).

7.4 ASSIGNMENT MODEL

Let us consider a large industrial public sector company dealing in the manufacture of electronic equipment involving a large number of components. The company, in order to promote ancillary industries, has decided to subcontract n of the components for which it has selected n outside contractors who are expected to submit an estimate for the manufacture of each of the n components. Each contractor can undertake the manufacture of one component only.

In a realistic situation, the contractors would have their own preferences in the choice of component. Consequently, their bids for the manufacture of those components in which they have no interest would be very high. The problem before the management of the public sector is to so allot the n contracts for the n components that the total cost incurred is minimal.

7.4.1 Mathematical Formulation

Let $i = 1, 2, \ldots, n$ represent the task to be performed and $j = 1, 2, \ldots, n$, the agents performing the task. Since each contractor can commit to only one task,

$$\begin{aligned} x_{ij} &= 1 \quad \text{if task } i \text{ is performed by } j \\ &= 0 \quad \text{otherwise.} \end{aligned}$$

Let c_{ij} be the corresponding cost. The optimization problem is:

$$\text{Minimize } Z = \sum_{i=1}^{n} \sum_{j=1}^{n} c_{ij}x_{ij}$$

subject to

$$\sum_{j=1}^{n} x_{ij} = 1 \quad \text{for } j = 1, 2, \ldots, n,$$

$$\sum_{j=1}^{n} x_{ij} = 1 \quad \text{for } i = 1, 2, \ldots, n,$$

$$x_{ij} = 1 \text{ or } 0 \quad \text{for all } i \text{ and } j.$$

The meaning of the first constraint is

$$\sum_{j=1}^{n} x_{ij} = x_{1j} + x_{2j} + \ldots + x_{nj}.$$

A particular contractor j is allowed to perform only one of the tasks $i = 1, 2, \ldots, n$. Hence, of the terms in the foregoing sequence, only one is non-zero. Similarly,

$$\sum_{j=1}^{n} x_{ij} = x_{i1} + x_{i2} + \ldots + x_{in}.$$

Since a particular task i is performed by one contractor only, in the foregoing sequence only one of the terms is non-zero.

The assignment problem can be easily posed as a transportation problem. Table 7.38 shows this explicitly.

TABLE 7.38

		Contractor				
		D_1	D_2	D_3	...	D_n
	T_1	C_{11}	C_{12}	C_{13}	...	C_{1n}
	T_2	C_{21}	C_{22}	C_{23}	...	C_{2n}
Task	T_3	C_{31}	C_{32}	C_{33}	...	C_{3n}
	⋮	⋮	⋮	⋮		⋮
	T_n	C_{n1}	C_{n2}	C_{n3}	...	C_{nn}

Table 7.38 gives the cost structure. The demand-availability pattern is as shown in Table 7.39.

The transportation model is of a particular type: the availabilities at the supply stations are all unity and the demands at an equal number of stations or depots are also unity.

TABLE 7.39

		Contractor				
		D_1	D_2	D_3	...	D_n
		1	1	1	...	1
	T_1 1					
	T_2 1					
Task	T_3 1					
	⋮ ⋮					
	T_n 1					

7.5 LINEAR PRODUCTION MODEL

7.5.1 Production Model

In a production activity, the problem is one of determining the optimum combination of input levels to achieve a particular objective. If all the input quantities are reduced to a common scale of value, such as cost, then the problem is to minimize cost at a fixed level of production or to maximize production for a fixed total budget. Assuming a fixed technological process, we find that the input factors could be combined in different proportions to obtain different outputs. The relationship between a firm's inputs and its outputs is called the *production function*. Most real production situations do not involve a simple choice among alternative input combinations for a *fixed process*. They involve *different processes*, each with its own possible combinations of inputs and outputs. These alternative processes might be alternative methods, procedures, schedules, or sequences of operations. Such alternative processes are called *activities*, and their output quantities, *activity levels*.

7.5.2 Linear Production Model

In a linear production model, three basic assumptions are involved. These are:

(i) The relationships among the variables are fixed over the planning period envisaged; that is, the model is static and not dynamic in time.

(ii) An activity requires a fixed proportion of input factors per unit of output, regardless of the total output level. For example, if a unit output requires 5 man-hours of labour and 20 units of material, then 100 units of output require 500 man-hours and 2000 units of material.

(iii) The activity levels are additive. This means that a number of activities can be carried out simultaneously, without affecting the fixed input proportions for each activity. If several activities have the same output products, then the total output from the joint activities is the sum of the outputs from the individual activities.

Let us denote the outputs by $x_1, x_2, \ldots, x_n$ and the input quantities consumed by $b_1, b_2, \ldots, b_m$. For a linear model, we have

$$
\begin{aligned}
a_{11}x_1 + a_{12}x_2 + \ldots + a_{1n}x_n &= b_1 \\
a_{21}x_1 + a_{22}x_2 + \ldots + a_{2n}x_n &= b_2 \\
&\vdots \\
a_{m1}x_1 + a_{m2}x_2 + \ldots + a_{mn}x_n &= b_m
\end{aligned}
$$

In these expressions, the constants a_{ij}'s are called the *technical coefficients of production.* The term a_{ij} denotes the input factor number i per unit of product output number j. For example, a_{11} is the amount of input number 1 per unit of product output number 1. If x_1 is the total output of product number 1, the total input of number 1 is $a_{11}x_1$. Similarly, a_{34} denotes the rate of input number 3 per unit of product output number 4. The given set of equations represents the *technological* relations between the inputs and the outputs.

If any of the inputs are restricted, because of their scarcity (as in the case of machine-hours or labour available), etc., these are termed *economic* restrictions. For example, input b_3 may be restricted to a maximum value of b_3^*. Then

$$b_3 \leqslant b_3^*.$$

In certain situations, the *output quantities* may be specified in advance. The problem in such a case may be to determine the least-cost combination of inputs that gives a desired amount of output. Then the restriction will be

$$x_j = x_j^*.$$

Or, the market survey might have revealed the *minimum* expected demand for a product x_j so that we have

$$x_j \geqslant x_j^*.$$

Similarly, if the *maximum* expected demand for a product j is x_j^*, we have the inequality

$$x_j \leqslant x_j^*.$$

Now, let us see how a linear production model can give rise to a linear programming problem. Let us assume that the same output x can be produced by a finite number s of alternative processes or activities and that these processes can be applied simultaneously (as we had assumed earlier for a linear production model). Let $\lambda_1, \lambda_2, \ldots, \lambda_s$ be the quantities of the product planned to be produced by the respective processes. The model now becomes

$$\begin{aligned}
x &= \lambda_1 + \lambda_2 + \ldots + \lambda_s \\
b_1 &= a_{11}\lambda_1 + a_{12}\lambda_2 + \ldots + a_{1s}\lambda_s \\
b_2 &= a_{21}\lambda_1 + a_{22}\lambda_2 + \ldots + a_{2s}\lambda_s \\
&\vdots \\
b_m &= a_{m1}\lambda_1 + a_{m2}\lambda_2 + \ldots + a_{ms}\lambda_s
\end{aligned} \tag{7.12}$$

If there are any economic restrictions or capacity limitations on, for instance, the inputs b_g and b_h, then we have

$$\begin{aligned}
b_g &\geqslant a_{g1}\lambda_1 + a_{g2}\lambda_2 + \ldots + a_{gs}\lambda_s, \\
b_h &\geqslant a_{h1}\lambda_1 + a_{h2}\lambda_2 + \ldots + a_{hs}\lambda_s.
\end{aligned} \tag{7.13}$$

If we assume that there is an infinite number of possible processes, $s = \infty$, it means that for a given x continuous (infinitesimal) variation of input proportion is possible. In other words, the production is continuous and differentiable, i.e.,

$$x = f(b_1, b_2, \ldots, b_m).$$

7.5.3 Linear Production Model for More Than One Product

It is possible to extend the foregoing formulation to an industry engaged in producing more than one product. Let us assume that the company is producing three products, 1, 2, and 3, and that each product is produced in s number of processes, involving m number of factors. Let the production levels for these products be x_1, x_2, and x_3. Then we have

$$\begin{aligned}
x_1 &= \beta_{11}\lambda_1 + \beta_{12}\lambda_2 + \ldots + \beta_{1s}\lambda_s, \\
x_2 &= \beta_{21}\lambda_1 + \beta_{22}\lambda_2 + \ldots + \beta_{2s}\lambda_s, \\
x_3 &= \beta_{31}\lambda_1 + \beta_{32}\lambda_2 + \ldots + \beta_{3s}\lambda_s,
\end{aligned} \tag{7.14}$$

$$\begin{aligned}
b_1 &= a_{11}\lambda_1 + a_{12}\lambda_2 + \ldots + a_{1s}\lambda_s \\
b_2 &= a_{21}\lambda_1 + a_{22}\lambda_2 + \ldots + a_{2s}\lambda_s \\
&\vdots \\
b_m &= a_{m1}\lambda_1 + a_{m2}\lambda_2 + \ldots + a_{ms}\lambda_s
\end{aligned} \tag{7.15}$$

In Eqs. (7.14) and (7.15), λ_1 is the level of activity number 1, λ_2 the level of activity number 2, and so on. When $\lambda_1 = 1$, the input number 1 consumed in activity (or process) number 1 is a_{11}, the input number 2 consumed in activity number 1 is a_{21}, and so on. For this activity level of 1, the number of units of the outputs 1, 2, and 3 are β_{11}, β_{21}, and β_{31}, respectively. The coefficients a_{ij}'s and β_{ij}'s are called input and output coefficients, respectively. In a practical situation, however, all s processes may not be involved in the output of each product. For example, if each process is involved in one output product only, then we have for the present case

$$s = 3,$$

$$\beta_{12} = \beta_{13} = \beta_{21} = \beta_{23} = \beta_{31} = \beta_{32} \equiv 0.$$

In addition, we can incorporate economic restrictions or capacity limitations as in inequalities (7.13).

7.5.4 Profit Maximization and Cost Minimization

After the production model has been expressed in a mathematical form, various processes can be combined in several ways, defining a set of feasible alternatives with respect to the s's. From among these sets of feasible solutions, we need to find an optimal solution. Generally, the optimality condition is to *maximize the profit*. When the total profit can be expressed as a linear production model, we have a problem in linear programming.

Let us consider the model as expressed by sets (7.12) and (7.13). Let p be the price at which the product is sold so that, if x is the total output, the amount obtained is

$$px = p(\lambda_1 + \lambda_2 + \ldots + \lambda_s).$$

Let the cost per unit input number 1 be q_1, the cost per unit input number 2 be q_2, and so on. Then the total cost of all the inputs used in producing outputs $\lambda_1, \lambda_2, \ldots, \lambda_s$ is

$$q_1(a_{11}\lambda_1 + a_{12}\lambda_2 + \ldots + a_{1s}\lambda_s) + q_2(a_{21}\lambda_1 + a_{22}\lambda_2 + \ldots + a_{2s}\lambda_s) + \ldots + q_m(a_{m1}\lambda_1 + a_{m2}\lambda_2 + \ldots + a_{ms}\lambda_s). \quad (7.16)$$

Hence, the gross profit earned is

$$(p - a_{11}q_1 - a_{21}q_2 - \ldots - a_{m1}q_m)\lambda_1 + (p - a_{12}q_1 - a_{22}q_2 - \ldots - a_{m2}q_m)\lambda_2 + \ldots + (p - a_{1s}q_1 - a_{2s}q_2 - \ldots - a_{ms}q_m)\lambda_s$$
$$= c_1\lambda_1 + c_2\lambda_2 + \ldots + c_s\lambda_s.$$

In the process of calculating this gross profit, we do not include those a_{ij}'s that form a part of the capacity restrictions, such as the ones given in inequalities (7.13). The problem therefore is to maximize the function

$$f = c_1\lambda_1 + c_2\lambda_2 + \ldots + c_s\lambda_s.$$

If, on the other hand, the problem is to find the least-cost combination of inputs by which a given number of products or a given sum of products is to be made, we have to minimize the quantity expressed by inequalities (7.16), i.e., to minimize

$$(q_1a_{11} + q_2a_{21} + \ldots + q_ma_{m1})\lambda_1 + (q_1a_{12} + q_2a_{22} + \ldots + q_ma_{m2})\lambda_2 + \ldots + (q_1a_{1s} + q_2a_{2s} + \ldots + q_ma_{ms})\lambda_s.$$

To make the foregoing points more specific, let us consider Problem 7.5.

Problem 7.5 **A company manufactures three different products, 1, 2, and 3. Each product needs to be processed through two departments, A and B. Department A has two machines, A_1 and A_2, and department B has three machines, B_1, B_2, and B_3. Product 1 can be made with either machine of type A or any of type B. Product 2 can be made with A_1 or any type B machine. Product 3 can be made with only A_2 or B_2. The time taken in minutes by one unit of each product on each type of machine is given in Table 7.40. The total available machine time per week for each machine, the cost per week of running the machine, the cost of raw material needed per unit product, and the selling cost per unit product are also given.**

TABLE 7.40

Machine	*Product* 1	2	3	*Available Time per Week (minutes)*	*Cost per Week at Full Capacity*
A_1	6	12	10	5000	250
A_2	8			8000	300
B_1	7	9		4000	275
B_2	5	8	7	6000	350
B_3	8	7		5000	200
Material Cost	0.2	0.3	0.5		
Selling Price	1.2	1.8	2		

We observe that product 1 can be made by six different processes, (A_1, B_1), (A_1, B_2), (A_1, B_3), (A_2, B_1), (A_2, B_2), or (A_2, B_3). Let $x_1, x_2, \ldots, x_6$ be the numbers of product 1 made per week by each of these processes. Product 2 can be made by any of the three processes (A_1, B_1), (A_1, B_2), and (A_1, B_3). Let x_7, x_8, and x_9, respectively, be the number of product 2 made per week by these three processes. We have only one way of making product 3, namely, (A_1, B_2); let x_{10} be the number of this product produced per week. In all, there are 10 different processes or activities for the manufacture of products 1, 2, and 3.

Now, to take the machine capacity restriction, we proceed as follows. The total time demand on machine A_1 in making product 1 is $(6x_1 + 6x_2 + 6x_3)$. The total time demand on A_1 for product 2 is $(12x_7 + 12x_8 + 12x_9)$, and for 3, it is $10x_{10}$. Hence, the total time demand on machine A_1 is

$$6x_1 + 6x_2 + 6x_3 + 12x_7 + 12x_8 + 12x_9 + 10x_{10}$$

and this should not exceed 5000 minutes.

Calculating similarly for other machines, we obtain the constraints

$$6x_1 + 6x_2 + 6x_3 + 12x_7 + 12x_8 + 12x_9 + 10x_{10} \leqslant 5000,$$
$$8x_4 + 8x_5 + 8x_6 \leqslant 8000,$$
$$7x_1 + 7x_4 + 9x_7 \leqslant 6000,$$
$$5x_2 + 5x_5 + 8x_8 + 7x_{10} \leqslant 6000,$$
$$8x_3 + 8x_6 + 7x_9 \leqslant 5000.$$

The next step is to arrive at the objective function. Product 1 can be made by six different processes. For each of these processes, we shall find the cost of manufacturing one unit of product 1, add the material cost, and subtract this sum from the selling price to get the profit per unit. The unit production cost of product 1 by process 1, i.e., (A_1, B_1), is

$$(6/5000)(250) + (7/4000)(275) = 0.3 + 0.48 = 0.78.$$

The material cost is 0.2, thus giving the total variable cost per unit as $0.2 + 0.78 = 0.98$. The selling price being 1.2, the unit profit is $c_1 = 0.22$. Similarly, we have

product 1 by $(A_1, B_2) = 0.3 + 0.29 = 0.59$,

unit profit $c_2 = 1.2 - (0.59 + 0.20) = 0.41$,

product 1 by $(A_1, B_3) = 0.3 + 0.32 = 0.62$,

unit profit $c_3 = 1.2 - (0.62 + 0.20) = 0.38$,

product 1 by $(A_2, B_1) = 0.3 + 0.48 = 0.78$,

unit profit $c_4 = 1.2 - (0.78 + 0.20) = 0.22$,

product 1 by $(A_2, B_2) = 0.3 + 0.29 = 0.59$,

unit profit $c_5 = 1.2 - (0.59 + 0.20) = 0.41$,

product 1 by $(A_2, B_3) = 0.3 + 0.31 = 0.61$,

unit profit $c_6 = 1.2 - (0.61 + 0.20) = 0.39$,

product 2 by $(A_1, B_1) = 0.6 + 0.62 = 1.22$,

unit profit $c_7 = 1.8 - (1.22 + 0.30) = 0.28$,

product 2 by $(A_1, B_2) = 0.6 + 0.47 = 1.07$,

unit profit $c_8 = 1.8 - (1.07 + 0.30) = 0.43$,

product 2 by $(A_1, B_3) = 0.6 + 0.28 = 0.88$,

unit profit $c_9 = 1.8 - (0.88 + 0.30) = 0.62$,

product 3 by $(A_1, B_2) = 0.50 + 0.41 = 0.91$,

unit profit $c_{10} = 2 - (0.91 + 0.5) = 0.59$.

The objective function is therefore (with appropriate values for c's substituted)

$$Z = c_1x_1 + c_2x_2 + c_3x_3 + \ldots + c_9x_9 + c_{10}x_{10}$$

and the problem is to maximize it. This shows how the linear programming technique can be applied to an industrial problem dealing with *optimal product mix and activity levels.* The solution provides the loading on each type of machine, the total amount of each product to be produced, and the fraction of each product produced by each of the possible processes or activities.

The solution to a problem of the foregoing type does not in general give integer values to the variables $x_1, x_2, \ldots, x_{10}$. However, as the production process is a continuous operation, the closest integer to each of these decision variables can be taken as the production level per week.

7.6 CRITICAL PATH SCHEDULING

7.6.1 Project Scheduling

Let us consider a small project concerning the organization of a conference. The project involves several preparatory activities before the conference can actually begin. These activities and the time taken to complete them are:

(a) Write to members for suitable date	6 days
(b) Inform members of conference date	2 days
(c) Prepare agenda	3 days
(d) Mail agenda and other relevant papers to members	4 days
(e) Arrange conference room	2 days
(f) Arrange meals for members	1 day
(g) Travel time required by members	6 days

As can be observed, all these activities cannot commence at the same time as some of them can start only after the completion of others. The activity that has to be completed before the beginning of another activity is called the *predecessor activity.* The activity that comes after the completion of another activity is called the *successor activity.* A few of the activities can be taken up simultaneously; these are called *concurrent activities.*

A chain of activities, such as the one just described, can be represented by a network as shown in Fig. 7.3. Such a network consists of nodes marked 1, 2, 3, . . . , and connected by arrows. The node represents the beginning or completion of an activity, and the arrows denote the activities themselves. The beginning or the starting (S) node is represented by 0, and the end or the terminal (T) node by 6. The duration taken by the various activities is indicated above the arrows. Node 0 represents the actual commencement of the conference. Any project, such as the one being discussed, need not result in a unique network representation. The network may change, depending on which activities are considered concurrent activities, which activities precede other activities, etc.

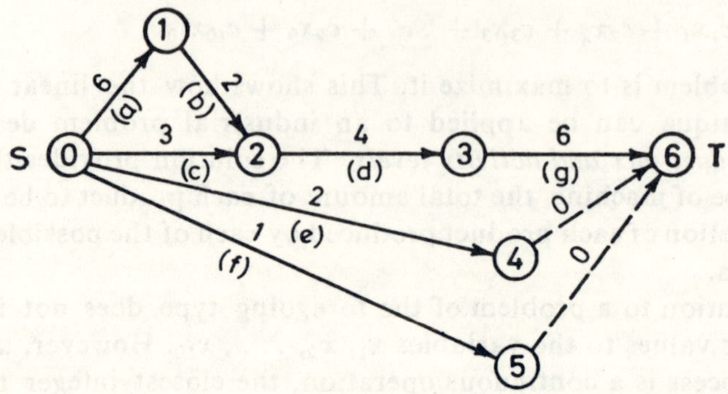

FIGURE 7.3

In Fig. 7.3, activity 0-1 represents activity (*a*), which is "write to members for suitable date"; activity 1-2 represents activity (*b*); and so on. Activity 2-3 cannot begin until activity 1-2 is completed. There are some activities, such as 4-5 and 5-6, which are shown by broken arrows. These stand for activities that consume zero time, but it is necessary that activities 0-4 and 0-5 be completed before node 6 is said to have occurred. Equivalently, we can say that nodes 4 and 5 should occur before node 6.

As remarked earlier, we can draw the network in different ways. For example, it is possible to merge nodes 4 and 6 without changing the problem or the meaning of the network. We have deliberately separated nodes 4 and 6 in order to bring into the picture the concept of *dummy* activities or arrows such as 4-6 and 5-6.

The problem now is to determine the minimum time required to start the conference, counting time from the occurrence of node S. The minimum time obviously is *not* the sum of the durations of all the activities, nor is it equal to the duration of the activity consuming the longest time. Take, for example, activity 2-3. This activity can begin only after activity 0-2 is completed, i.e., after 3 days. Further, activity 2-3 cannot begin until activities 0-1 and 1-2 are also completed; and these two activities take up a total of 8 days. Hence, we conclude that activity 2-3 can begin only 8 days after the occurrence of node 0. In this way, we can find the *earliest occurrence time* for each node in the network. We shall start assessing the duration with the occurrence of node S at time zero:

node S is 0

node 1 is 6

node 2 (based on path 0-1-2) is 8

node 3 is 12

node 4 is 2

node 5 is 1

node (6) T (based on path 2-3-4) is 18

Hence, the minimum time required before the conference can actually begin is 18 days. In this process, we are led to determining the length of the longest chain between the two points S and T.

7.6.2 Solution by Linear Programming

In order to solve this problem by the techniques of linear programming, we proceed as follows. We imagine that a traveller walks from 0 to 6 along a succession of links. Each link is indicated by x_{ij} connecting node i to node j. If the traveller uses link x_{ij}, then x_{ij} will be given a value 1; if not, it will be zero. The traveller comes to a particular node j via one of the paths, for example, x_{ij}, and leaves it by another path, say, x_{jk}. So, if a node j has several arrows pointing towards it, for instance, gj, hj,

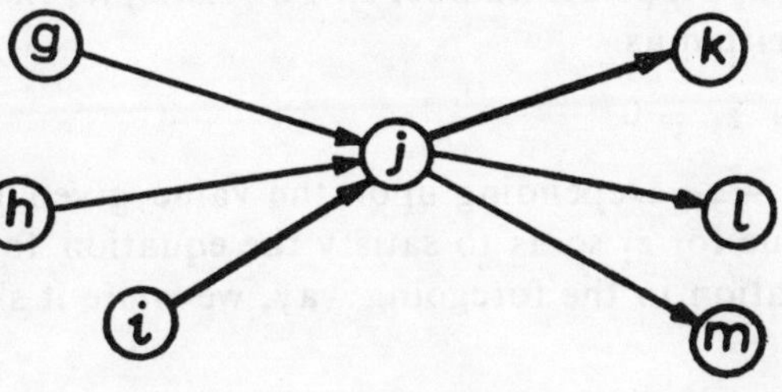

FIGURE 7.4

and ij (Fig. 7.4), and several arrows pointing away from it, for instance, jk, jl, and jm, then we must have, for node j, the equation

$$x_{gj} + x_{hj} + x_{ij} - x_{jk} - x_{jl} - x_{jm} = 0.$$

This is because, for each of the links x_{ij} and x_{jk}, the value is 1 (the traveller uses this path), whereas for each of the links—x_{gj}, x_{hj}, x_{jl}, and x_{jm}—the value is zero (the traveller does not use any of these paths). For the network shown in Fig. 7.4, we get

$$\begin{aligned} &x_{01} - x_{12} = 0,\\ &x_{12} + x_{02} - x_{23} = 0,\\ &x_{23} - x_{36} = 0, \\ &x_{04} - x_{46} = 0,\\ &x_{05} - x_{56} = 0. \end{aligned} \tag{7.17}$$

Further, since the traveller has to start from S along one of the paths leading from S, we have

$$-x_{01} - x_{02} - x_{04} - x_{05} = -1. \tag{7.18}$$

Similarly, as the traveller has to reach T along one of the paths, we get

$$x_{36} + x_{46} + x_{56} = 1. \tag{7.19}$$

Now, we can consider that the time taken to reach a particular node j from node i is equal to the time taken for that particular activity, namely, ij. Hence, the length of the longest chain from S to T is obtained from:

$$\text{Maximum } Z = 6x_{01} + 3x_{02} + 2x_{04} + x_{05} + 2x_{12} + 4x_{23} + 6x_{36} + 0.x_{46} + 0.x_{56}. \tag{7.20}$$

This is a linear programming problem with a special structure. Every variable x_{ij} appears once with a positive coefficient (+1), and once with a negative coefficient (−1). This problem can be made to agree with the transportation model, provided we take care of the negative coefficients appearing in Eqs. (7.17). To do this, we proceed as follows. For a vari able, or a sum of variables appearing with a negative sign in any of the constraint equations, we introduce the difference between a large constant M and an appropriate number z_i. For example, the first equation in (7.17) can be written as

$$x_{01} - M + z_1 = 0,$$

where $x_{12} = M - z_1$. Depending upon the value given to M, we get an appropriate value for z_1 so as to satisfy the equation for x_{12}. Instead of writing the equation in the foregoing way, we write it slightly differently as

$$x_{12} + z_1 = M, \qquad x_{01} + z_1 = M.$$

The constant M is made large enough so that the same constant (with different z's) can be used for all the other equations also. Using this, we write Eqs. (7.17) as

$$\begin{aligned} x_{12} + z_1 &= M, & x_{01} + z_1 &= M, \\ x_{23} + z_2 &= M, & x_{12} + x_{02} + z_2 &= M, \\ x_{36} + z_3 &= M, & x_{23} + z_3 &= M, \\ x_{46} + z_4 &= M, & x_{04} + z_4 &= M, \\ x_{56} + z_5 &= M, & x_{05} + z_5 &= M. \end{aligned} \tag{7.21}$$

In addition to these, we have two more equations, (7.18) and (7.19). Repeating them here, we get

$$\begin{aligned} x_{01} + x_{02} + x_{04} + x_{05} &= 1, \\ x_{36} + x_{46} + x_{56} &= 1. \end{aligned} \tag{7.22}$$

These equations are now in the form of a transportation model. Since we have to maximize the objective function Z, we can consider this maximizing as equivalent to the maximization of the profit. Equations (7.21) and (7.22) can be considered the rim requirements, i.e., the plant capacities and sales requirements. The variables x_{ij} can assume a value zero or $(+1)$. The transportation matrix is as shown in Table 7.41. All the columns and

TABLE 7.41

	M	M	M	M	M	1
M	z_1	x_{12}				
M		z_2	x_{23}			
M			z_3			x_{36}
M				z_4		x_{46}
M					z_5	x_{56}
1	x_{01}	x_{02}		x_{04}	x_{05}	

rows add up to their marginal values. The profit index for each cell would be as follows:

For x_{ij}'s:	their coefficients in the objective function
For z_i's:	zero
For all unoccupied cells:	a very high negative value so that they will not be filled in

Following the matrix minimum method, discussed in Section 7.2.3, we first fill in the cell with the maximum profit index. There are two cells, one corresponding to x_{01} and the other to x_{36}, both with the same index 6. Choosing one first and then the other, we can continue to fill in the table. The result is as shown in Table 7.42.

TABLE 7.42

	M	M	M	M	M	1
M	$M-1$	1				
M		$M-1$	1			
M			$M-1$			1
M				M		0
M					M	0
1	1					

To see whether this is the optimal solution, we compute the fictitious profits as shown in Table 7.43, with the true profit index entered in the occupied cells.

TABLE 7.43

							f.c.
	0	2					0
		0	4				−2
			0			6	−6
				0		0	−12
					0	0	−12
	6						6
f.c.	0	2	6	12	12	12	12

The fictitious profits are now calculated and entered in the unoccupied cells of the x_{ij}'s. There are only three such cells and they are shown in Table 7.44. The fictitious profits are higher than the true profit indices.

TABLE 7.44

							f.c.
							0
							−2
							−6
							−12
							−12
		8		18	18		6
f.c.	0	2	6	12	12	12	

Hence, the scheme shown in Table 7.44 is the optimal solution. According to this table, the variables x_{ij} with the value 1 are x_{01}, x_{12}, x_{23}, and x_{36}. The longest chain or the path traversed is 0-1-2-3-6 (Fig. 7.3). The total length (or time) is

$$6 + 2 + 4 + 6 = 18 \text{ days.}$$

This is the *critical path* connecting S and T. This path is critical in the sense that it consumes the maximum time in going from S to T. Any delay in the execution of the activities lying along this path will change the project duration. Extreme care must be taken to see that these critical activities are completed within the stipulated periods. The activities not lying along the critical path are not crucially affected by delays. For example, activity 0-4 can start even 16 days after the occurrence of S. Similarly, activity 0-2 can begin even 5 days after the commencement of the project.

7.6.3 Application of Duality Theorem

The solution for the critical path scheduling can be obtained more easily by the application of the duality theorem, discussed in Chapter 6. We shall state the primal again so that the dual can be written on the guidelines suggested in Chapter 6.

Primal

$$\text{Maximize } Z = 6x_{01} + 3x_{02} + 2x_{04} + x_{05} + 2x_{12} + 4x_{23} + 6x_{36}$$

subject to

$$\begin{aligned}
&-x_{01} - x_{02} - x_{04} - x_{05} = -1,\\
&x_{01} - x_{12} = 0,\\
&x_{02} + x_{12} - x_{23} = 0,\\
&x_{23} - x_{36} = 0,\\
&x_{04} - x_{46} = 0,\\
&x_{05} - x_{56} = 0,\\
&x_{36} + x_{46} + x_{56} = 1,\\
&x_{ij} = 1 \text{ or } 0.
\end{aligned} \tag{7.23}$$

The dual of this can be written as:

$$\text{Minimize } w = -y_0 + y_6$$

subject to

$$\begin{aligned}
&-y_0 + y_1 \geqslant 6,\\
&-y_0 + y_2 \geqslant 3,\\
&-y_0 + y_4 \geqslant 2,\\
&-y_0 + y_5 \geqslant 1,\\
&-y_1 + y_2 \geqslant 2,\\
&-y_2 + y_3 \geqslant 4,\\
&-y_3 + y_6 \geqslant 6,\\
&-y_4 + y_6 \geqslant 0,\\
&-y_5 + y_6 \geqslant 0,\\
&-\infty < y_i < \infty.
\end{aligned} \tag{7.24}$$

Since the objective function is merely the difference between two variables, y_0 and y_6, we can arbitrarily choose $y_0 = 0$ and solve for the rest. Thus, we get

$$y_1 \geqslant 6, \qquad y_2 \geqslant 3, \qquad y_4 \geqslant 2, \qquad y_5 \geqslant 1,$$

$$y_2 \geqslant 2 + y_1, \qquad y_3 \geqslant 4 + y_2,$$

$$y_6 \geqslant 6 + y_3, \qquad y_6 \geqslant y_4, \qquad y_6 \geqslant y_5.$$

Since we are trying to minimize $w = y_6$, we look for the algebraical minimum value of y_6. Thus, we get

$$y_1 \geqslant 6, \tag{7.25}$$

$$y_2 \geqslant 3, \text{ also } y_2 \geqslant 2 + y_1,$$

and hence

$$y_2 \geqslant 8; \tag{7.26}$$

$$y_3 \geqslant 4 + y_2,$$

and hence

$$y_3 \geqslant 12; \tag{7.27}$$

$$y_6 \geqslant 6 + y_3,$$

and hence

$$y_6 \geqslant 18;$$

$$y_6 \geqslant y_4, \qquad y_4 \geqslant 2,$$

and hence

$$y_6 \geqslant 2;$$

$$y_6 \geqslant y_5, \qquad y_5 \geqslant 1,$$

and hence

$$y_6 \geqslant 1.$$

The minimum value for y_6 is therefore 18. Since $y_6 \geqslant 6 + y_3$, we get

$$18 \geqslant 6 + y_3 \quad \text{or} \quad y_3 \leqslant 12.$$

This combined with inequality (7.27) gives

$$y_3 = 12.$$

Since $y_3 \geqslant 4 + y_2$, we get

$$12 \geqslant 4 + y_2 \quad \text{or} \quad y_2 \leqslant 8.$$

This combined with inequality (7.26) gives

$$y_2 = 8.$$

Since $y_2 \geqslant 2 + y_1$, we get

$$8 \geqslant 2 + y_1 \quad \text{or} \quad y_1 \leqslant 6.$$

This combined with inequality (7.25) gives

$$y_1 = 6.$$

The final values are therefore

$$\begin{aligned} &y_1^* = 6, \quad y_2^* = 8, \quad y_3^* = 12, \\ &18 \geqslant y_4^* \geqslant 2, \quad 18 \geqslant y_5^* \geqslant 1, \quad y_6^* = 18. \end{aligned} \tag{7.28}$$

The objective function is therefore $w^* = 18$, the same value as obtained for the objective function Z in the primal. The advantage of using the duality theorem is obvious, since the objective function in this case consists of only two variables. To identify the critical path in the original network, we make use of the complementary slackness theorem, discussed in Section 6.3.

The solution given in Eqs. (7.28) indicates that the 2nd, 3rd, 4th, 8th, and 9th constraints in (7.24) are satisfied as inequalities. This means that the slack variables appearing in these constraints are non-zero. Consequently, the corresponding coefficients in the primal, i.e., x_{02}, x_{04}, x_{05}, x_{46}, and x_{56}, are zero. Hence, the primal solutions are

$$(x_{ij}) = (1, 0, 0, 0, 1, 1, 1, 0, 0).$$

The critical path is therefore $(x_{01} - x_{12} - x_{23} - x_{36})$.

7.7 ALLOCATION PROBLEM

7.7.1 Job Allocation

The problem we shall now discuss is generally known as the *allocation problem.* Let us assume that there are n number of workers and n number of tasks to be performed. Each worker can be engaged in the execution of any of these n jobs, but the number of units of tasks completed per week depends on the nature of the task and the particular workman performing it. For example, let us consider the case where there are three workers, A_1, A_2, and A_3, and three different *types* of jobs, J_1, J_2, and J_3, to be performed. If A_1 is engaged in jobs of type J_1 alone, he can turn out 1 unit of this type per week. If he is assigned to jobs of type J_2, he can produce 2 units of them per week, whereas if he is put on type J_3, he can turn out 3 units of them per week. These are the productivity levels of A_1 with respect to jobs J_1, J_2, and J_3. The productivities of A_1 and those of the other two workers can be expressed as in Table 7.45.

The problem is to assign to each worker a particular type of task that will make the total number of jobs done per week by all the workers a maximum.

TABLE 7.45

		Task		
		J_1	J_2	J_3
	A_1	1	2	3
Workman	A_2	2	4	1
	A_3	3	1	5

Problem 7.6 A firm is interested in bringing out four types of optical equipment. There are four different workshops where these can be produced. Because of the differences in the nature of facilities available in these workshops and the varying skills of their workmen, the production capacities of the workshops vary, depending on the type of optical units they manufacture. These productivities can be expressed as in Table 7.46. It is clear from this table that if workshop A_1 is engaged in

TABLE 7.46

		Optical Equipment			
		O_1	O_2	O_3	O_4
	A_1	1	1	2	4
Workshop	A_2	2	3	1	1
	A_3	4	1	2	3
	A_4	1	3	3	2

the production of equipment O_1, it can produce one unit per day; if it is assigned the production of equipment O_4, it can produce four units per day. Similarly, if workshop A_2 is allotted the production of equipment type O_2, it can produce three units per week, and so on. The problem is to allocate to each workshop the task of producing a particular optical unit such that the total number of optical units produced per week is a maximum.

Such problems are generally known as *allocation problems.* They can be solved by the Simplex method, but a special algorithm is available, which is less cumbersome. However, we shall use the Simplex method to solve this allocation problem.

Problem 7.7 Let us consider the case of the three workers and the three types of tasks performed by them stated in Section 7.7.1. As the productivities are the same as those expressed in Table 7.45, we repeat the table here.

TABLE 7.45

		Task		
		J_1	J_2	J_3
	A_1	1	2	3
Workman	A_2	2	4	1
	A_3	3	1	5

We introduce a variable x_{ij}, which takes on a value of either 1 or 0. If worker i is engaged in task j, then x_{ij} assumes a value 1; if not, the value will be 0. For example, if worker 2 is engaged in task 2, then $x_{21} = 0$, $x_{22} = 1$, and $x_{23} = 0$. Similarly, if $x_{33} = 1$, it means that worker 3 is engaged in task 3, and $x_{31} = x_{32} = 0$. This is equivalent to saying that a worker is engaged in any *one particular task*, and his productivities, as far as the remaining two tasks are concerned, are 0.

The productivity of worker 1 is therefore $x_{11} + 2x_{12} + 3x_{13}$. We note that only one of the variables x_{11}, x_{12}, and x_{13} will assume a value 1, depending on the particular task (1, 2, or 3) that worker 1 is allocated. If he is engaged in task 2, then $x_{12} = 1$, and his productivity will be 2; if he is engaged in task 3, then $x_{13} = 1$, and his productivity will be 3; and so on. The total productivity of all the three workers is therefore

$$Z = x_{11} + 2x_{12} + 3x_{13} + 2x_{21} + 4x_{22} + x_{23} + 3x_{31} + x_{32} + 5x_{33}.$$

Since each worker is engaged in only one type of task, the constraints expressing this will be

$$x_{11} + x_{12} + x_{13} = 1,$$
$$x_{21} + x_{22} + x_{23} = 1,$$
$$x_{31} + x_{32} + x_{33} = 1.$$

Further, since a particular type of task is performed by only one worker, the constraints expressing this will be

$$x_{11} + x_{21} + x_{31} = 1,$$
$$x_{12} + x_{22} + x_{32} = 1,$$
$$x_{13} + x_{23} + x_{33} = 1,$$
$$x_{ij} = 1 \text{ or } 0.$$

Subject to these constraints, the problem is to maximize Z.

This linear programming problem can be solved by the usual Simplex method The final scheme is given in Table 7.47. The last row shows that

TABLE 7.47

	x_{11}	x_{12}	x_{13}	x_{21}	x_{22}	x_{23}	x_{31}	x_{32}	x_{33}	Z	
x_{13}	1	0	1	1	0	1	0	−1	0	0	1
x_{22}	0	0	0	1	1	1	0	0	0	0	1
x_{31}	1	0	0	1	0	0	1	0	0	0	1
x_{33}	−1	0	0	−1	0	0	0	1	1	0	0
x_{12}	0	1	0	−1	0	−1	0	1	0	0	0
	0	0	0	−1	0	−4	0	−3	0	−1	−10

all the coefficients are negative, which means (in a maximization problem such as this one) that we have obtained the optimal solution. The maximum value for Z is 10, with $x_{13} = 1$, $x_{22} = 1$, and $x_{31} = 1$. The remaining x_{ij}'s are zero. This means that workman 1 will be assigned type J_3 job, workman 2 type J_2, and workman 3 type J_1.

Further, in the last row for Z, there is an additional zero corresponding to the x_{11} column. This means that the solution we have obtained is not a unique solution. As a matter of fact, $x_{11} = 1$, $x_{22} = 1$, and $x_{33} = 1$, with the other x_{ij}'s being equal to zero, also yield $Z = 10$.

EXERCISES

1. A company has three warehouses, A, B, and C, located in three different cities. There are four distribution centres, a, b, c, and d, to which the product manufactured by the company is to be shipped from the three warehouses. The costs of shipping one unit from the warehouses to the four distribution centres, the capacities of the warehouses, and the requirements of centres a, b, c, and d are as tabulated here. Determine

		Capacity	*Shipping Cost to Centre*			
			a	b	c	d
	A	100	3	5	7	11
Warehouse	B	130	1	4	6	3
	C	170	5	8	12	7
Requirement			150	120	80	50

how the shipping should take place in order to minimize the shipping cost.
[*Ans.* a: 130 units from B, 20 units from C; b: 20 units from A, 100 units from C; c: 80 units from A; d: 50 units from C; shipping cost: 2040 monetary units.]

2. In the task of taking over wholesale dealing in wheat trade, the management is faced with a distribution problem. There are three central collection centres, A, B, and C, located in three different cities in a particular zone. Due to conditions such as storage capacity, administrative limitations, and payment to the farmers, the capacities, in tonnes, of these three centres are restricted respectively to 90,000, 75,000, and 35,000 per season. From these three centres, wheat has to be shipped to five distribution centres, a, b, c, d, and e, for distribution to retailers. The demands, in tonnes, at these distribution centres are 36,000, 31,000, 62,000, 27,000, and 44,000, respectively. The demand balances the supply. There is no particular vested interest to meet the demand of any one particular distribution centre from a particular depot. The transportation costs per tonne from the collection centres A, B, and C to the distribution centres a, b, c, d, and e are as tabulated here. What quantity of

		Distribution Centre				
		a	b	c	d	e
Collection Centre	A	1.5	6.4	1.8	4	3.5
	B	1.6	2.6	1.9	3.1	5.8
	C	5.3	3.5	2.4	1.3	2.2

wheat should the management distribute to each centre so that its transportation cost is minimized?
[*Ans.*

		a	b	c	d	e
(in thousands of tonnes)	A	36		18		36
	B		31	44		
	C				27	8

The total shipping cost is 410,300 units.]

3. Consider Exercise 2. The capacities in tonnes of the three main collection centres are the same, namely, 90,000, 75,000, and 35,000. The demands of the five distribution centres however are now:

a	b	c	d	e
40,000	35,000	70,000	30,000	50,000

The demand now exceeds the supply. Determine the distribution pattern purely from the economic standpoint, i.e., the pattern that gives the minimum cost of transportation.

[*Ans.*

		a	b	c	d	e
(in thousands of tonnes)	A	40		30		20
	B		35	40		
	C				30	5

The total shipping cost is 40,100 monetary units. *Note*: Satisfying the economic standpoint alone, the distribution pattern is unfair to centre e since its demand is met only 50%, whereas the requirements of the other centres are fully met!]

4. Use the fictitious cost method to solve the problem of transporting goods from the sources A, B, and C to the destinations D, E, F, and G when the demand-availability is as indicated by the marginal values tabulated here. The other values in the table denote the cost of transportation per unit between stations.

		D	E	F	G
		4	8	3	6
A	9	5	3	1	9
B	4	5	7	3	2
C	9	3	2	1	8

[*Ans.*

	4	8	3	6	P	u_i
9		$(3+x)$	3	$(2-x)$	1	0
4				4		−7
9	4	$(5-x)$		x		−1
v_j	4	3	1	9	0	

$0 \leqslant x \leqslant 2$; cost: 60 units.]

5. An ore supplier who owns two plants, A and B, enters into a contract with three processing plants, C, D and E, to supply each of his plants 20 tonnes of ore per week. As per this contract, a customer who does not receive all he requires is entitled to a penalty cost of 150 monetary units per tonne of shortage. The costs of transportation per unit tonne are as tabulated here. During a particular period, the production capacities of plants A and B have come down to 30 tonnes per week

	C	D	E
A	15	35	25
B	10	50	40

and 20 tonnes per week, respectively. What should be the distribution pattern for the ore supplier (a) if he cannot obtain additional ore from other producers? (b) if he can buy additional ore from another producer at 100 monetary units per tonne?
Ans. (a)

	C	D	E	u_i
A	0	10	20	0
B	20			−5
S		10		115
v_i	15	35	25	

6. The cost structure and the demand-availability in a transportation problem are as tabulated here. Find the least-cost solution.

	20	80	20	40
90	5	2	1	9
20	1	2	3	2
90	3	2	1	8

[*Ans.*

	20	80	20	40	40	u_i
90		$10 + x$	20	$20 - x$	40	0
20				20		−7
90	20	$70 - x$		x		−1
v_i	4	3	1	9	0	

$0 \leqslant x \leqslant 20$; cost: 470.]

7. Solve the problem of transporting material from the factories A, B, C, and D to the depots L, M, N, and P when the demand-availability and the cost structure are as tabulated here.

		Depot			
		L	M	N	P
		9	15	21	10
Factory	A 15	18	15	6	15
	B 21	5	7	5	6
	C 9	21	23	10	25
	D 13	12	16	2	18

Ans.

	L	M	N	P
A		12		
B	8	3		10
C			9	
D	1		12	

8. Solve the following trans-shipment problem. The cost structure is:

	D_1	D_2	D_3	D_4	S_1	S_2	S_3
S_1	5	4	3	2	0	2	1
S_2	10	8	4	7	1	0	4
S_3	9	9	8	4	3	2	0
D_1	0	1	3	2	5	9	9
D_2	2	0	2	3	4	6	7
D_3	2	3	0	1	3	4	9
D_4	4	1	2	0	4	7	3

The supply-demand structure is:

		D_1	D_2	D_3	D_4
		1	4	4	6
S_1	5				
S_2	5				
S_3	5				

Ans.

	D_1	D_2	D_3	D_4	S_1
S_1	1			7	
S_2			2		3
S_3				5	
D_4		6			

9. Solve the assignment problem

	1	1	1	1	1	1
1	9	22	58	11	19	27
1	43	78	72	50	63	48
1	41	28	91	37	45	33
1	74	42	27	49	39	32
1	36	11	57	22	25	18
1	3	56	53	31	17	28

Ans.

			1		
					1
	1				
		1			
				1	
1					

8

Parametric Linear Programming

8.1 INTRODUCTION

Consider a linear programming problem cast in a standard form as discussed in Section 2.2:

Determine $x_1 \geqslant 0, x_2 \geqslant 0, \ldots, x_n \geqslant 0$

so as to

minimize $Z = c_1x_1 + c_2x_2 + \ldots + c_nx_n$

subject to

$$a_{11}x_1 + a_{12}x_2 + \ldots + a_{1n}x_n = b_1$$
$$\vdots$$
$$a_{m1}x_1 + a_{m2}x_2 + \ldots + a_{mn}x_n = b_m$$

In the analysis of the problems discussed so far, it has been explicitly assumed that the cost coefficients $c_1, c_2, \ldots$, the stipulations $b_1, b_2, \ldots$, and the structural coefficients $a_{11}, a_{12}, \ldots$ are fixed quantities. It is easy to visualize that, in many practical situations, these coefficients are not fixed, but vary. For example, if the objective function Z represents the cost of producing a product, then the cost coefficients $c_1, c_2, \ldots$ vary from day to day or week to week. Similarly, if the stipulations $b_1, b_2, \ldots$ represent, for example, the time available for the use of different machines, a more realistic statement would be to give limits or a range for each b_i as b_i' and b_i'', i.e.,

$$b_i' \leqslant b_i \leqslant b_i''.$$

When all the coefficients are assumed to be random variables with appropriate distribution functions associated with each coefficient, the problem becomes a *stochastic* linear programming problem.

In this chapter, we shall consider problems that are more restricted. Here only the cost coefficients are allowed to vary. Further, the variation of c_i is of the form

$$c_i = c_i' + \lambda c\ .$$

The cost coefficients depend on a parameter λ which can take any finite value. The problem remains linear in form, but, because of the introduction of the parameter λ, it is named a *parametric linear programming* problem. Instead of a single parameter λ, the cost coefficients can depend on two parameters, u and v, as

$$c_i = c_i' + c_i''u + c_i'''v,$$

where we can find the optimum values for the objective function for specific combinations of the parameters u and v.

8.2 PROCEDURE FOR SOLVING PARAMETRIC LINEAR PROGRAMMING PROBLEMS

The general method for the solution of parametric linear programming problems can best be illustrated through an example.

Problem 8.1

$$\text{Maximize } Z = (6 - \lambda)x_1 + (12 - \lambda)x_2 + (4 - \lambda)x_3$$

subject to

$$3x_1 + 4x_2 + x_3 \leqslant 2,$$
$$x_1 + 3x_2 + 2x_3 \leqslant 1,$$
$$x_1, x_2, x_3 \geqslant 0.$$

When introducing the slack variables x_4 and x_5, Tableau I appears as shown. If the coefficients in the last row remain positive for a certain

TABLEAU I

Basis	x_1	x_2	x_3	x_4	x_5	
x_4	3	4	1	1	0	2
x_5	1	3	2	0	1	1
	$(\lambda - 6)$	$(\lambda - 12)$ ↑	$(\lambda - 4)$			0

value of λ, then the solution in Tableau I is optimal for that value of λ. To determine this, we observe that

$(\lambda - 6)$ is positive for $\lambda > 6$,

$(\lambda - 12)$ is positive for $\lambda > 12$,

$(\lambda - 4)$ is positive for $\lambda > 4$.

Hence, for all values of $\lambda \geqslant 12$, the solution in Tableau I is optimal. If λ becomes less than 12, the first coefficient in the last row to become negative is $(\lambda - 12)$. Hence, we bring x_2 into the basis. According to the usual criterion, x_2 replaces x_5. The corresponding tableau then appears as now shown.

TABLEAU II

Basis	x_1	x_2	x_3	x_4	x_5	
x_4	5/3	0	−5/3	1	−4/3	2/3
x_2	1/3	1	2/3	0	1/3	1/3
	$(\frac{2}{3}\lambda - 2)$	0	$(\frac{1}{3}\lambda + 4)$	0	$(-\frac{1}{3}\lambda + 4)$	$-\frac{1}{3}\lambda + 4$

Considering the coefficients in the last row of Tableau II, we find that

$(\frac{2}{3}\lambda - 2)$ is positive for $\lambda > 3$,

$(\frac{1}{3}\lambda + 4)$ is positive for $\lambda > -12$,

$(-\frac{1}{3}\lambda + 4)$ is positive for $\lambda > 12$.

Hence, the ranges shown in Fig. 8.1 indicate that, for the values of λ between 12 and 3, the solution in Tableau II is optimal.

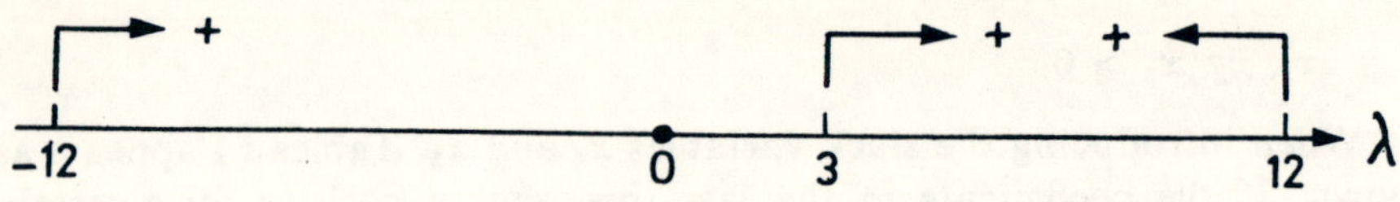

FIGURE 8.1

When λ becomes less than 3, the coefficient $(\frac{2}{3}\lambda - 2)$ becomes negative first. Hence, x_1 replaces x_4 in the basis. Tableau III is the result.

TABLEAU III

Basis	x_1	x_2	x_3	x_4	x_5	
x_1	1	0	−1	3/5	−4/5	2/5
x_2	0	1	1	−1/5	3/5	1/5
	0	0	$(\lambda + 2)$	$(-\frac{2}{5}\lambda + \frac{6}{5})$	$(\frac{1}{5}\lambda + \frac{12}{5})$	$\frac{24}{5} - \frac{3}{5}\lambda$

For the coefficients in the last row of Tableau III, we see that

$(\lambda + 2)$ is positive for $\lambda > -2$,

$(-\frac{2}{5}\lambda + \frac{6}{5})$ is positive for $\lambda < 3$,

$(\frac{1}{5}\lambda + \frac{12}{5})$ is positive for $\lambda > -12$.

All the coefficients will therefore be positive if λ lies between 3 and -2, as shown in Fig. 8.2. It can be observed that $\lambda = 0$ lies in this range and,

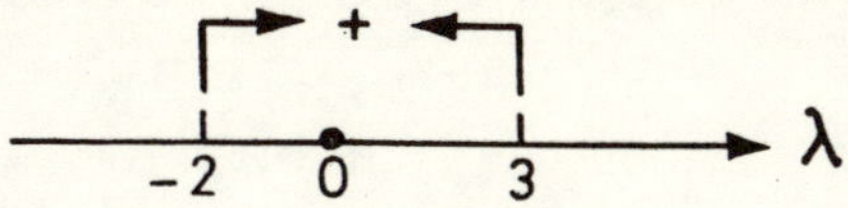

FIGURE 8.2

for $\lambda = 0$, the problem is a standard (non-parametric) linear programming problem with the solutions $x_1 = 2/5$, $x_2 = 1/5$, $x_3 = 0$, and $Z_{max} = 24/5$, as seen from Tableau III. When λ becomes less than 2, x_3 replaces x_2 in the basis. The result is given in Tableau IV.

TABLEAU IV

Basis	x_1	x_2	x_3	x_4	x_5	
x_1	1	1	0	2/5	$-1/5$	3/5
x_3	0	1	1	$-1/5$	3/5	1/5
	0	$(-\lambda - 2)$	0	$(-\frac{1}{5}\lambda + \frac{8}{5})$	$(-\frac{2}{5}\lambda + \frac{6}{5})$	$\frac{22}{5} - \frac{4}{5}\lambda$

It should be observed from Tableau IV that

$(-\lambda - 2)$ is positive for $\lambda < -2$,

$(-\frac{1}{5}\lambda + \frac{8}{5})$ is positive for $\lambda < 8$,

$(-\frac{2}{5}\lambda + \frac{6}{5})$ is positive for $\lambda < 3$.

Hence, all the coefficients will be positive if λ does not exceed -2. Thus, Tableaux I–IV cover all the ranges for the parameter λ. The ranges for λ and the corresponding solutions are

$\lambda \geqslant 12$, $Z_{max} = 0$ with $x_1 = x_2 = x_3 = 0$,

$3 \leqslant \lambda \leqslant 12$, $Z_{max} = (4 - \frac{1}{3}\lambda)$ with $x_1 = 0$, $x_2 = \frac{1}{3}$, $x_3 = 0$,

$-2 \leqslant \lambda \leqslant 3$, $Z_{max} = (\frac{24}{5} - \frac{3}{5}\lambda)$ with $x_1 = \frac{2}{5}$, $x_2 = \frac{1}{5}$, $x_3 = 0$,

$\lambda \leqslant -2$, $Z_{max} = (\frac{22}{5} - \frac{4}{5}\lambda)$ with $x_1 = \frac{3}{5}$, $x_2 = 0$, $x_3 = \frac{1}{5}$.

These are shown in Fig. 8.3.

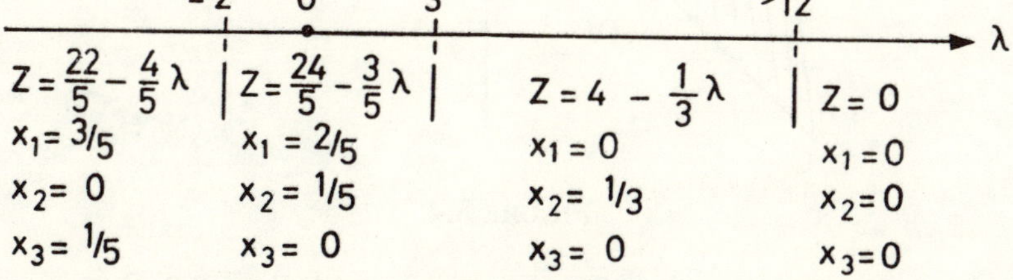

FIGURE 8.3

Problem 8.2

Maximize $Z = (2 + 2u + v)x_1 + (1 + u - v)x_2$

subject to

$x_1 + x_2 \leqslant 3,$

$2x_1 - x_2 \leqslant 2,$

$x_1, x_2 \geqslant 0.$

In this problem, there are two parameters u and v. We shall consider them together. Consider Tableau I where x_3 and x_4 are slack variables.

TABLEAU I

Basis	x_1	x_2	x_3	x_4	
x_3	1	1	1	0	3
x_4	2	-1	0	1	2
	$-(2 + 2u + v)$	$-(1 + u - v)$			0

For the coefficients in the last row of Tableau I,

$-2 - (2u + v)$ is positive for $2u + v \leqslant -2$,

$-1 - (u - v)$ is positive for $u - v \leqslant -1$.

Figure 8.4 shows the region (labelled I) enclosed by these two limits. Line

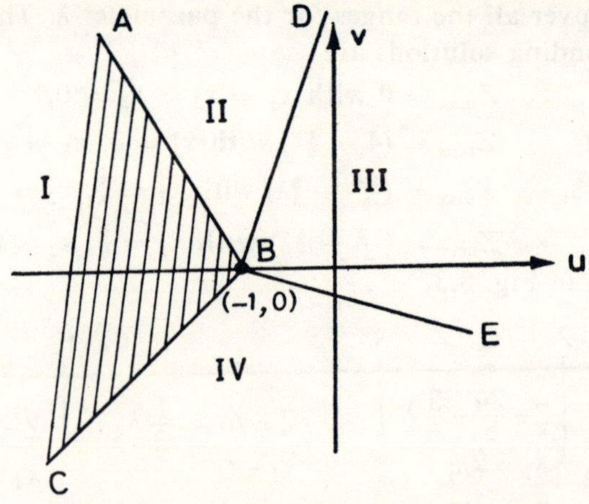

FIGURE 8.4

AB represents the equation $2u + v = -2$, and line BC the equation $u - v = -1$. Region I contains the points (u, v) satisfying both the constraints

$$2u + v \leqslant -2,$$

$$u - v \leqslant -1.$$

In this region, the values given in Tableau I are optimal, with $Z_{max} = 0$, $x_1 = 0$, and $x_2 = 0$.

Let us explore the region to the right of AB. Here x_1 replaces x_4 in the basis. The corresponding values are shown in Tableau II.

TABLEAU II

Basis	x_1	x_2	x_3	x_4	
x_3	0	3/2	1	−1/2	2
x_1	1	−1/2	0	1/2	1
	0	$(-2 - 2u + \frac{1}{2}v)$	0	$(1 + u + \frac{1}{2}v)$	$2 + 2u + v$

Analyzing the coefficients in Tableau II, we find that

$(-2 - 2u + \frac{1}{2}v)$ is positive for $-2u + \frac{1}{2}v \geqslant 2$,

$(1 + u + \frac{1}{2}v)$ is positive for $u + \frac{1}{2}v \geqslant -1$.

The two conditions may be written as $-4u + v \geqslant 4$ and $2u + v \geqslant -2$. The lines corresponding to the equations $-4u + v = 4$ and $2u + v = -2$ are BD and BA and both pass through the point B (−1, 0). In the region (labelled II in Fig. 8.4) enclosed between lines BA and BD, the values of Tableau II are optimal.

Next, consider the region to the right of BD. This means introducing x_2 to replace x_3 in the basis. Tableau III gives the values.

TABLEAU III

Basis	x_1	x_2	x_3	x_4	
x_2	0	1	2/3	−1/3	4/3
x_1	1	0	1/3	1/3	5/3
	0	0	$\frac{1}{3}(4 + 4u - v)$	$\frac{1}{3}(1 + u + 2v)$	$\frac{1}{3}(14 + 14u + v)$

For the coefficients in the last row of Tableau III,

$\frac{1}{3}(4 + 4u - v)$ is positive if $4u - v \geqslant -4$,

$\frac{1}{3}(1 + u + 2v)$ is positive if $u + 2v \geqslant -1$.

Region III represents these constraints and, for this region, Tableau III gives the optimal values.

Tableau IV gives the results when x_4 replaces x_1 in the basis.

TABLEAU IV

Basis	x_1	x_2	x_3	x_4	
x_2	1	1	1	0	3
x_4	3	0	1	1	5
	$(-1-u-2v)$	0	$(1+u-v)$	0	$(3+3u-v)$

Again,

$$(-1-u-2v) \text{ is positive for } -u-2v \geqslant 1,$$

$$(1+u-v) \text{ is positive for } u-v \geqslant -1.$$

The region corresponding to these two constraints is between lines BC and BE in Fig. 8.4 and is marked IV. Hence, the final results are

Region ABC

$$2u+v \leqslant -2,$$

$$u-v \leqslant -1,$$

$$Z_{max} = 0, \quad x_1 = 0, \quad x_2 = 0;$$

Region ABD

$$2u+v \geqslant -2,$$

$$4u-v \leqslant -4,$$

$$Z_{max} = 2+2u+v, \quad x_1 = 1, \quad x_2 = 0;$$

Region DBE

$$4u-v \geqslant -4,$$

$$u+2v \geqslant -1,$$

$$Z_{max} = \tfrac{1}{3}(14+14u+v), \quad x_1 = \tfrac{5}{3}, \quad x_2 = \tfrac{4}{3};$$

Region CBE

$$u+2v \leqslant -1,$$

$$u-v \geqslant -1,$$

$$Z_{max} = 3+3u-v, \quad x_1 = 0, \quad x_2 = 3.$$

EXERCISES

1. Minimize $Z = -x_1 + x_2 + \lambda y$ for all values of λ
subject to

$$3x_1 + x_4 + 2x_5 = 12,$$

$$3x_2 - x_4 + x_5 + y = 3,$$

$$x_3 + x_4 + x_5 = 9,$$

$$x_1, x_2, x_3, x_4, x_5, y \geqslant 0.$$

[*Ans.* For $\lambda \geqslant \frac{1}{3}$, $Z_{\min} = -3$; for $\frac{1}{3} \geqslant \lambda \geqslant -\frac{1}{3}$, $Z_{\min} = -4 + 3\lambda$; for $\lambda \leqslant -\frac{1}{3}$, $Z_{\min} = -1 + 12\lambda$.]

2. Maximize $Z = (3 - \lambda)x_1 + (6 - \lambda)x_2 + (2 - \lambda)x_3$ for all values of λ subject to the constraints

$$3x_1 + 4x_2 + x_3 \leqslant -2,$$

$$x_1 + 3x_2 + 2x_3 \leqslant -1,$$

$$x_1, x_2, x_3 \geqslant 0.$$

[*Ans.* For $\lambda \geqslant 6$, $Z_{\max} = 0$; for $6 \geqslant \lambda \geqslant \frac{3}{2}$, $Z_{\max} = 2 - \frac{1}{3}\lambda$; for $\lambda \leqslant \frac{3}{2}$, $Z_{\max} = \frac{1}{5}(12 - 3\lambda)$.]

3. Minimize $Z = -x_1 + 2x_2$
subject to the constraints

$$\lambda x_1 + x_2 \leqslant 2,$$

$$3x_1 + 2x_2 \leqslant 5,$$

$$x_1, x_2 \geqslant 0$$

for all values of λ.
[*Ans.* For $\lambda \geqslant \frac{6}{5}$, we have $x_1 = 2/\lambda$, $x_2 = 0$; for $\lambda \leqslant \frac{6}{5}$, we have $x_1 = 5/3$, $x_2 = 0$.]

4. Minimize $8y_1 + 3y_2 + 2y_3$
subject to the constraints

$$6y_1 - y_2 + 3y_3 \geqslant 6 + u,$$

$$3y_1 + 5y_2 + 2y_3 \geqslant 5 + v,$$

$$y_1, y_2, y_3 \geqslant 0$$

for all values of u and v.
[*Ans.* For

$$5u + v \geqslant -35, \qquad 2u - 3v \leqslant 3,$$

we have

$$y_1 = 0, \qquad y_2 = (3 - 2u + 3v)/17, \qquad y_3 = (79 + 4u + 11v)/17;$$

for

$$2u - 3v \geqslant 3, \qquad u \geqslant -6,$$

we have

$$y_1 = 0, \qquad y_2 = 0, \qquad y_3 = 2 + \frac{u}{3};$$

for

$$u \leqslant -6, \qquad v \leqslant -5,$$

we have

$$y_1 = y_2 = y_3 = 0;$$

for

$$v \geqslant -5, \qquad 5u + v \leqslant -35,$$

we have

$$y_1 = y_3 = 0, \qquad y_2 = 1 + \frac{v}{5}.]$$

9

Integer Programming

9.1 INTRODUCTION

In the problems that we have discussed so far, the variables $x_1, x_2, \ldots$ were subjected to non-negativity constraints only. Consequently, in the process of obtaining an optimal solution, these variables could assume any positive values—integer or fractional. However, in practice, situations occur where the variables cannot assume fractional values. For example, if the variables $x_1, x_2, \ldots$ represent numbers of different machines such as lathes, drilling machines, and grinders, then an optimization problem involving these machines should provide only an integer-valued solution. Solutions such as 2.76 lathes and 4.3 drilling machines do not have any meaning. Programming problems where the variables are allowed to assume only non-negative integer values are called *integer programming problems*. If all the variables $x_1, x_2, \ldots, x_n$ are constrained to be integer-valued, the problem is then called a *pure integer programming problem*. If only some variables are constrained to be integer-valued, then it is called a *mixed integer programming problem*. The transportation and trans-shipment problems belong to the integer programming class. However, because of the particular nature of these problems, special algorithms were developed to solve them. In this chapter, we shall study only pure integer programming problems.

9.2 CUTTING-PLANE METHOD

The method described here is applicable to pure integer programming problems, but can be easily extended to mixed integer programming problems also. Let the pure integer programming problem be stated as:

$$\text{Maximize } Z = c_1x_1 + c_2x_2 + \ldots + c_nx_n, \tag{9.1}$$

subject to

$$\begin{aligned} a_{11}x_1 + a_{12}x_2 + \ldots + a_{1n}x_n &= b_1 \\ a_{21}x_1 + a_{22}x_2 + \ldots + a_{2n}x_n &= b_2 \\ &\vdots \\ a_{m1}x_1 + a_{m2}x_2 + \ldots + a_{mn}x_n &= b_m \end{aligned} \tag{9.2}$$

$$x_1, x_2, \ldots, x_n, \text{ all non-negative integers.} \tag{9.3}$$

In the cutting-plane method, the problem is first solved without taking into account that the variables are to be integers. If the optimal solution to this problem gives all the integer values to the variables, then the problem is solved completely. However, in the optimal solution to the modified problem, one or more basic variables appear with non-integer values; then we solve the original problem with an additional constraint.

Let us assume that, for the modified problem, the final result in the Simplex method appears as shown in Tableau I. Let us assume also that

TABLEAU I

Basis	x_1	x_2	$\ldots$	x_k	$\ldots$	x_m	x_{m+1}	$\ldots$	x_n	Z
x_1	1						$u_{1,m+1}$	$\ldots$	$u_{1,n}$	v_{10}
x_2		1					$u_{2,m+1}$	$\ldots$	$u_{2,n}$	v_{20}
$\vdots$										
x_k				1			$u_{k,m+1}$	$\ldots$	$u_{k,n}$	v_{k0}
$\vdots$										
x_m						1	$u_{m,m+1}$	$\ldots$	$u_{m,n}$	v_{m0}
							$Z_{0,m+1}$	$\ldots$	$Z_{0,n}$	Z_{00}

v_k, the value of the basis x_k, is not an integer. In the row k, splitting all the constants corresponding to the non-basic variables into their largest lower integers and non-negative fractions, we get

$$\begin{aligned} u_{k,m+1} &= N_{k,m+1} + f_{k,m+1} \qquad 0 \leqslant f_{k,m+1} \leqslant 1 \\ u_{k,m+2} &= N_{k,m+2} + f_{k,m+2} \qquad 0 \leqslant f_{k,m+2} \leqslant 1 \\ &\vdots \\ v_{k0} &= N_{k0} + f_{k0} \qquad 0 \leqslant f_{k0} \leqslant 1 \end{aligned} \tag{9.4}$$

Now, we form the product

$$f_{k,m+1}x_{m+1} + f_{k,m+2}x_{m+2} + \ldots + f_{k,n}x_n = g_k \geqslant 0. \tag{9.5}$$

It is necessary to remember that each row in the Simplex tableau represents an equation. For example, the equation corresponding to the k-th row is

$$x_k + u_{k,m+1}x_{m+1} + u_{k,m+2}x_{m+2} + \ldots + u_{k,n}x_n = v_{k0}.$$

Substituting from Eqs. (9.4) the integral and fractional parts for each of the constants, we get

$$\begin{aligned} x_k &+ (N_{k,m+1} + f_{k,m+1})x_{m+1} + (N_{k,m+2} + f_{k,m+2})x_{m+2} \\ &+ \ldots + (N_{k,n} + f_{k,n})x_n = N_{k0} + f_{k0}. \end{aligned} \tag{9.6}$$

Since $g_k \geqslant 0$ from Eq. (9.5), we can observe that

$$x_k + N_{k,m+1}x_{m+1} + N_{k,m+2}x_{m+2} + \ldots + N_{k,n}x_n \leqslant N_{k0} + f_{k0}.$$

Since the left-hand side quantity is an integer, we conclude that

$$x_k + N_{k,m+1}x_{m+1} + N_{k,m+2}x_{m+2} + \ldots + N_{k,n}x_n \leqslant N_{k0}. \tag{9.7}$$

Subtracting Eq. (9.6) from inequality (9.7), we have

$$-f_{k,m+1}x_{m+1} - f_{k,m+2}x_{m+2} - \ldots - f_{k,n}x_n \leqslant -f_{k0}. \tag{9.8}$$

Adding a slack variable, we can convert inequality (9.8) into an equality as

$$-f_{k,m+1}x_{m+1} - f_{k,m+2}x_{m+2} - \ldots - f_{k,n}x_n + x_{n+1} = -f_{k0}. \tag{9.9}$$

This equation is called the *cut* and is introduced as a new constraint. This constraint equation brings into the problem an additional variable, x_{n+1}.

In the new constraint equation, all the constants are fractional parts. To get an idea of a fractional part, let us observe the following:

$+3.5 = +3.0 + 0.5,$

$-3.5 = -4.0 + 0.5$ (−4 is the largest lower integral part),

$+\frac{10}{3} = +3.0 + \frac{1}{3}.$

Observe also

$-\frac{10}{3} = -4.0 + \frac{2}{3}.$

When Eq. (9.9) is introduced as the new constraint into the last Simplex tableau, the tableau will no longer be the last since the newly introduced basis x_{n+1} will have the negative value $-f_{k0}$. This demands continuation of the iterative operation. Let us consider the example that follows.

Problem 9.1

Maximize $Z = 21x_1 + 11x_2$

subject to

$7x_1 + 4x_2 \leqslant 13,$

x_1 and x_2 being non-negative integers.

The given constraint can be transformed into an equality:

$7x_1 + 4x_2 + x_3 = 13,$

x_1, x_2, and x_3 being non-negative integers. We find the optimal solution ignores the fact that the variables are integers. The Simplex tableau therefore appears as Tableau I.

TABLEAU I

Basis	x_1	x_2	x_3	Z	
x_3	7	4	1		13
	−21 ↑	−11		1	0

The final result is as shown in Tableau II. The optimal solution is

TABLEAU II

Basis	x_1	x_2	x_3	Z
x_1	1	4/7	1/7	13/7
	0	1	3	39

$Z = 39$ and $x_1 = 1\frac{6}{7}$. Since x_1 is not an integer, we now introduce a cut. The cut, according to constraint (9.9), involves the non-basic variables (x_2, x_3) and the fractional parts of their constants. Thus, the cut is

$$-\tfrac{4}{7}x_2 - \tfrac{1}{7}x_3 + x_4 = -\tfrac{6}{7}.$$

The new linear programming problem is now:

Maximize $Z = 21x_1 + 11x_2$

subject to

$$7x_1 + 4x_2 + x_3 = 13,$$

$$-\tfrac{4}{7}x_2 - \tfrac{1}{7}x_3 + x_4 = -\tfrac{6}{7},$$

x_1 and x_2 being non-negative integers.

Ignoring the restriction on non-integer values, we get the Simplex tableau (labelled Tableau I). The last row in Tableau I indicates that

TABLEAU I

Basis	x_1	x_2	x_3	x_4	Z
x_1	1	4/7	1/7		13/7
x_4		−4/7	−1/7	1	−6/7
	0	1	3	0	39

optimality has been reached, but x_4 as a basis is infeasible since its value is negative. So we can conveniently apply the Dual Simplex method. To determine the new basis, we derive the *ratios* as in Table 1. Since this is

RATIO TABLE 1

non-basic variable	x_2	x_3
coefficient in last row	1	3
coefficient in row 2	−4/7	−1/7
Ratio	−7/4	−7/3

a maximization problem, we choose the maximum ratio. Hence, x_2 enters as the new basis. The result is shown in Tableau II. Now, x_2 is fraction-

TABLEAU II

Basis	x_1	x_2	x_3	x_4	Z
x_1	1	0	0	1	1
x_2		1	1/4	−7/4	3/2
	0	0	11/4	7/4	$37\frac{1}{2}$

valued. The cut involves the non-basic variables (x_3, x_4) and their fractional parts. Thus, observing that the fractional part of $-\frac{7}{4}$ $(= -2 + \frac{1}{4})$ is $\frac{1}{4}$, we have

$$-\tfrac{1}{4}x_3 - \tfrac{1}{4}x_4 + x_5 = -\tfrac{1}{2}.$$

With this as the new constraint, the Simplex tableau appears as Tableau III.

TABLEAU III

Basis	x_1	x_2	x_3	x_4	x_5	Z
x_1	1	0	0	1		1
x_2		1	1/4	−7/4		3/2
x_5			−1/4	−1/4	1	−1/2
	0	0	11/4	7/4	0	75/2

To remove x_5 from the basis, the ratios in Table 2 are derived.

RATIO TABLE 2

non-basic variable	x_3	x_4
coefficient in last row	11/4	7/4
coefficient in row 3	−1/4	−1/4
Ratio	−11	−7

Choosing the maximum ratio, we obtain x_4 as the new basis. The result is then as in Tableau IV.

TABLEAU IV

Basis	x_1	x_2	x_3	x_4	x_5	Z
x_1	1	0	−1	0	4	−1
x_2		1	2	0	−7	5
x_4			1	1	−4	2
	0	0	1	0	7	34

Now, x_1 has become infeasible. The ratios are as in Table 3.

RATIO TABLE 3

non-basic variable	x_3	x_5
coefficient in last row	1	7
coefficient in row 1	−1	4
Ratio	*−1*	

With x_3 as the new basis, we get Tableau V. Now, all the basic vari-

TABLEAU V

Basis	x_1	x_2	x_3	x_4	x_5	Z
x_3	−1	0	1	0	−4	1
x_2	2	1	0	0	1	3
x_4	1	0	0	1	0	1
	1	0	0	0	11	33

ables are integer-valued and the optimal solution is $x_1 = 0$, $x_2 = 3$, and $Z = 33$.

The solution to the foregoing illustrative problem shows that the cutting-plane method involves the following steps.

Step 1 Find an optimal solution to the linear programming problem ignoring the integer stipulation but requiring that all the variables be non-negative.

Step 2 If the final tableau after Step 1 gives integer-valued solutions, the iteration is stopped. Otherwise, select a fractional-valued basic variable. If there are several fractional-valued basic variables, it is advisable to choose that basic variable which has the largest fractional part. With this, augment the current linear programming problem by an additional constraint in the form given by (9.9).

Step 3 Find the new optimal solution to the augmented problem and return to Step 2.

Whenever Step 2 is performed, a new slack variable is introduced. This tends to make the tableau bigger. In order to keep the linear programming problem from growing beyond bounds, we should adopt the following procedure: whenever the current optimal basic solution includes a slack variable introduced at a previous iteration, drop the row containing that slack variable in the current tableau before performing the next iteration.

Problem 9.2 A manufacturing concern produces three types of machine, A, B, and C, each of which makes use of two kinds of electronic sub-assembly units in different numbers. Let these two sub-assembly units be designated Unit 1 and Unit 2. Each type A machine requires four of Unit 1 and four of Unit 2. Each type B machine makes use of six of Unit 1 and three of Unit 2. The requirements of each type C machine are two of Unit 1 and five of Unit 2. The total number of Unit 1 and Unit 2 sub-assemblies available at any one time are 22 and 25, respectively. The profit factors associated with types A, B, and C machine are respectively 5, 6, and 4. Determine how many units of A, B, and C should be produced at a time in order to maximize the profit.

Mathematically, this problem may be stated as follows:

$$\text{Maximize } Z = 5x_1 + 6x_2 + 4x_3$$

subject to

$$4x_1 + 6x_2 + 2x_3 \leqslant 22,$$

$$4x_1 + 3x_2 + 5x_3 \leqslant 25,$$

x_1, x_2, and x_3 being non-negative integers.

Using slack variables, we get Tableau I.

TABLEAU I

Basis	x_1	x_2	x_3	x_4	x_5	Z
x_4	4	6	2	1		22
x_5	4	3	5		1	25
	−5	−6 ↑	−4			0

Using the Simplex criterion, we have in Tableau II x_2 as the basis instead of x_4.

TABLEAU II

Basis	x_1	x_2	x_3	x_4	x_5	Z
x_2	2/3	1	1/3	1/6		11/3
x_5	2	0	4	−1/2	1	1/4
	−1	0	−2 ↑	1	0	22

Bringing in x_3 as the new basis, we get Tableau III. The last row

TABLEAU III

Basis	x_1	x_2	x_3	x_4	x_5	Z
x_2	1/2	1	0	5/24	−1/12	5/2
x_3	1/2	0	1	−1/8	1/4	7/2
	0	0	0	3/4	1/2	29

shows that an optimal solution has been obtained, but both x_2 and x_3 are fractional-valued. Choosing x_2, we get the additional constraint to be introduced:

$$-\tfrac{1}{2}x_1 - \tfrac{5}{24}x_4 - \tfrac{11}{12}x_5 + x_6 = -\tfrac{1}{2}.$$

The corresponding tableau is now given (see Tableau IV).

TABLEAU IV

Basis	x_1	x_2	x_3	x_4	x_5	x_6	Z
x_2	1/2	1	0	5/24	−1/12		5/2
x_3	1/2	0	1	−1/8	1/4		7/2
x_6	−1/2	0	0	−5/24	−11/12	1	−1/2
	0	0	0	3/4	1/2		29

To remove x_6, we derive the ratios as in Table 1.

RATIO TABLE 1

non-basic variable	x_1	x_4	x_5
coefficient in last row	0	3/4	1/2
coefficient in row 3	−1/2	−5/24	−11/12
Ratio	*0*	−18/5	−6/11

Hence, x_1 enters as the new basis in Tableau V.

TABLEAU V

Basis	x_1	x_2	x_3	x_4	x_5	x_6	Z
x_2	0	1	0	0	−1	1	2
x_3	0	0	1	−1/3	−2/3	1	3
x_1	1	0	0	5/12	11/6	−2	1
	0	0	0	3/4	1/2		29

Tableau V gives the optimal solution, with $x_1 = 1$, $x_2 = 2$, $x_3 = 3$, and $Z = 29$. In formulating the additional constraint condition after Tableau III, we chose x_2. We could have chosen x_3 as well since both x_2 and x_3 had equal fractional parts. The constraint condition would then have been

$$-\tfrac{1}{2}x_1 - \tfrac{7}{8}x_4 - \tfrac{1}{4}x_5 + x_6 = -\tfrac{1}{2}.$$

EXERCISES

1. Maximize $Z = x_1 + 3x_2$
subject to the constraints

$$2x_1 - 3x_2 \leqslant 4,$$

$$-x_1 + 2x_2 \leqslant 7,$$

$$3x_1 + x_2 \leqslant 9,$$

x_1 and x_2 being non-negative integers.
[*Ans.* $Z = 13$, $x_1 = 1$, $x_2 = 4$.]

2. Maximize $Z = 2x_1 + 3x_2$
subject to the constraints

$$-x_1 + x_2 \leqslant 2,$$

$$47x_1 + 8x_2 \leqslant 188,$$

$$3x_1 + 2x_2 \leqslant 19,$$

x_1 and x_2 being non-negative integers.
[*Ans.* $Z = 11$, $x_1 = 3$, $x_2 = 5$.]

3. Minimize $Z = 3x_1 + 10x_2$
subject to the constraints

$$2x_1 + x_2 \geqslant 6,$$

$$-x_1 + x_2 \leqslant 8,$$

$$x_1 + 5x_2 \leqslant 12,$$

x_1 and x_2 being non-negative integers.
[*Ans.* $Z = 9$, $x_1 = 3$, $x_2 = 0$.]

4. Minimize $Z = 10x_1 + 14x_2 + 21x_3$
subject to the constraints

$$8x_1 + 11x_2 + 9x_3 \geqslant 12,$$

$$2x_1 + 2x_2 + 7x_3 \geqslant 14,$$

$$9x_1 + 6x_2 + 3x_3 \geqslant 10,$$

x_1, x_2, and x_3 being non-negative integers.
[*Ans.* $Z = 52$, $x_1 = 1$, $x_2 = 0$, $x_3 = 2$.]

10

Mathematical Aspects of Linear Programming

10.1 INTRODUCTION

So far, sufficient theoretical fundamentals of linear programming have been explained to provide an understanding and application of the techniques it employs to solve a variety of problems. However, a good insight can be obtained by considering certain generalized aspects of the linear programming problem. In this chapter, we shall consider these aspects in detail, using matrix notation for compactness and clarity. It is assumed that the reader knows the elements of matrix operations. However, some of the points discussed earlier will be repeated here so as to make this chapter more or less self-contained.

10.2 CONVEX SETS AND EXTREME POINTS

In linear programming, the word "set" is used to refer to a collection of points in an n-dimensional space. Just as (x_1, x_2, x_3) can be considered a point in a three-dimensional space, so too $X = (x_1, x_2, \ldots, x_n)$ can be imagined as a point in an n-dimensional space. In a three-dimensional space, the point (x_1, x_2, x_3) can be considered the terminal point of a vector joining the origin and the point (x_1, x_2, x_3). In such a case, the point (x_1, x_2, x_3) is called a vector in a three-dimensional vector space. Similarly, the point $X = (x_1, x_2, \ldots, x_n)$ can be called a vector in an n-dimensional vector space. Let us consider two points $U = (u_1, u_2, \ldots, u_n)$ and $V = (v_1, v_2, \ldots, v_n)$. A *segment* joining U and V is defined as the locus of all points X obtained by combining U and V as

$$X = aU + (1 - a)V, \tag{10.1}$$

where $0 \leqslant a \leqslant 1$.

A *convex set* is defined as a set of points that has the following property: a segment joining any two points in the set is also entirely in the set. By convention, we say a set consisting of only one point is convex. The geometrical meaning of a segment in a two-dimensional space is shown in Fig. 10.1. Vector OA is $\mathbf{V}$ and vector OB is $\mathbf{U}$. Vector AB is $(\mathbf{U} - \mathbf{V})$. Let C be an intermediate point such that the ratio of the segment AC to

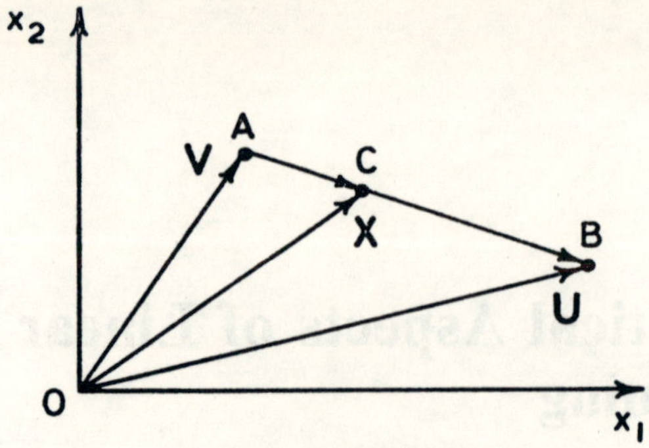

FIGURE 10.1

the length AB is $a(0 \leqslant a \leqslant 1)$. Vector AC is then $a(\mathbf{U} - \mathbf{V})$. Vector OC is $\mathbf{X}$. We therefore have

$$\text{vector } AC = \text{vector } OC - \text{vector } OA,$$

i.e.,

$$a(\mathbf{U} - \mathbf{V}) = \mathbf{X} - \mathbf{V}$$

or

$$\mathbf{X} = a\mathbf{U} + (1 - a)\mathbf{V}.$$

As a is varied from 0 to 1, point C moves from A to C.

In a two-dimensional space, a convex set and a non-convex set appear as in Fig. 10.2. Figures 10.2(a) and 10.2(b) are convex sets, whereas

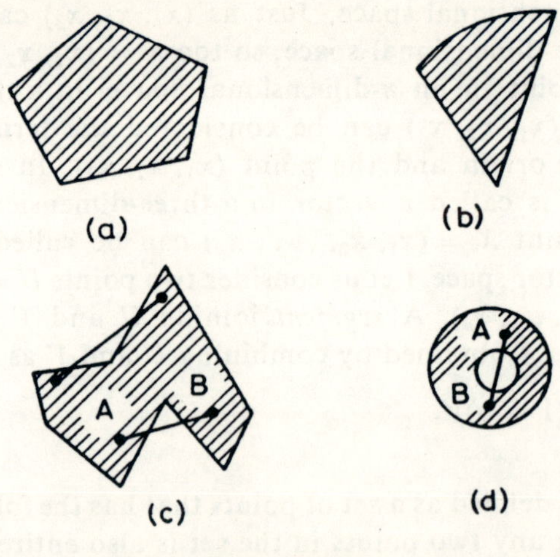

FIGURE 10.2

Figs. 10.2(c) and 10.2(d) are not. In Figs. 10.2(c) and 10.2(d), not all the points lying on the line joining A and B belong to the shaded regions.

An *extreme point* of a convex set is a point in the set that is not on any segment joining two other points in the set. In other words, X is an extreme point of a convex set if and only if there do not exist other points U, V, $U \neq V$, in the set such that $X = aU + (1 - a)V$, $0 < a < 1$. We should note the strict inequality imposed on a. An extreme point is a boundary point of the set; however, not all boundary points of a convex set are necessarily extreme points. In Fig. 10.2(a), the vertices are extreme points. Point X, which is a boundary point, is not an extreme point, because it can be represented as a convex combination of U and V, with $0 < a < 1$.

10.3 CONVEX COMBINATION

Convex combination is a special form of linear combination. Let $U, V, W, \ldots$ be n independent vectors and let X be a vector such that

$$X = a_1U + a_2V + a_3W + \ldots,$$

where all the a's are $\geqslant 0$ and their sum $a_1 + a_2 + a_3 + \ldots$ is equal to 1. Then X is said to be a convex combination of the vectors $U, V, W, \ldots$.

The significance of the convex combination in relation to the convex set is that, given a convex set S with the extreme points $E_1, E_2, \ldots, E_n$, all the points in the set S can be expressed as a convex combination of its extreme points. This can be illustrated by a simple example. For the convex set shown in Fig. 10.3, let E_1, E_2, E_3, and E_4 be the corner points,

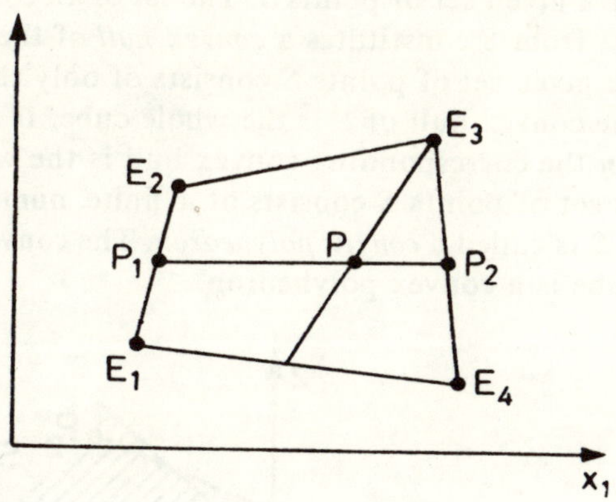

FIGURE 10.3

and let P be a point in the set. We can express P as a convex combination of the extreme points. To do this, draw a line P_1P_2 passing through P such that P_1 and P_2 are on the boundary. P is on the segment joining P_1 and P_2 and hence is a convex combination of P_1 and P_2, i.e.,

$$P = a_1P_1 + a_2P_2,$$

where $a_1, a_2 \geqslant 0$, $a_1 + a_2 = 1$. Further, P_1 and P_2 are on the segments joining E_1, E_2, and E_3, E_4. Hence,

$$P_1 = e_1E_1 + e_2E_2,$$

where $e_1, e_2 \geqslant 0$, $e_1 + e_2 = 1$;

$$P_2 = e_3E_3 + e_4E_4,$$

where $e_3, e_4 \geqslant 0$, $e_3 + e_4 = 1$. Substituting these, we get

$$\begin{aligned} P &= a_1(e_1E_1 + e_2E_2) + a_2(e_3E_3 + e_4E_4) \\ &= a_1e_1E_1 + a_1e_2E_2 + a_2e_3E_3 + a_2e_4E_4. \end{aligned}$$

Hence, P is a linear combination of the corner points E_1, E_2, E_3, and E_4.

It remains to be shown that the sum of the coefficients is equal to 1. We note that

$$a_1(e_1 + e_2) + a_2(e_3 + e_4) = a_1 + a_2 = 1.$$

Therefore, P is a convex combination of the four extreme points. We should carefully note that this is not the only convex combination for P. It can also be expressed as a convex combination of the extreme points E_1, E_3, and E_4, as shown in Fig. 10.3.

10.4 CONVEX HULL, CONVEX POLYHEDRON, CONVEX CONE, AND SIMPLEX

Let us consider a given set of points S. The set of all convex combinations of sets of points from S constitutes a *convex hull* of the given set S. For example, if the given set of points S consists of only the eight vertices of a cube, then the convex hull of S is the whole cube. If S is the boundary of a circle, then the corresponding convex hull is the whole circle.

If the given set of points S consists of a finite number of points, the convex hull of S is called a *convex polyhedron*. The convex hull of the eight vertices of a cube is a convex polyhedron.

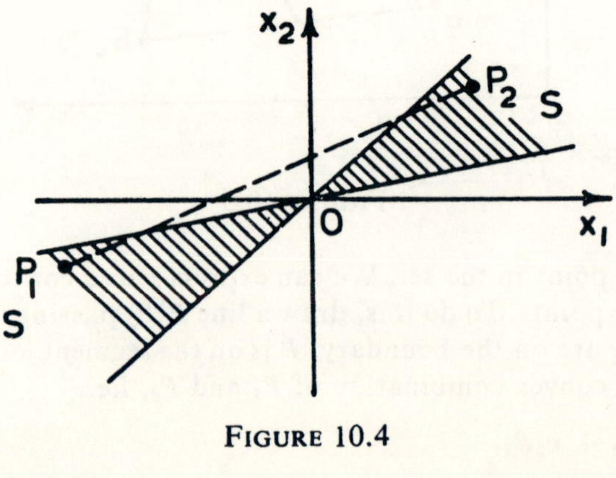

FIGURE 10.4

A set of vectors S is called a *cone* if, for every vector X in S, aX is also in S, where a is a non-negative number. The whole space is an example of a cone. The set S shown in Fig. 10.4 is a cone.

A *convex cone* is a cone that is convex. The cone shown in Fig. 10.4 is not a convex cone, because if we take the two points P_1 and P_2, then the segment joining them is not in S. That part of S lying in the first quadrant is a convex cone.

A *Simplex* is an n-dimensional convex polyhedron having exactly $(n + 1)$ vertices. A Simplex in zero dimension is a point; in one dimension it is a line, in two dimensions it is a triangle, and in three dimensions it is a tetrahedron.

10.5 LINEAR PROGRAMMING PROBLEM

The general linear programming problem is to find a vector $X = (x_1, x_2, \ldots, x_n)$ that minimizes an objective function $cX = (c_1x_1 + c_2x_2 + \ldots + c_nx_n)$, which is of a linear form, subject to the linear constraints $AX = b$ and $X \geqslant 0$, where $A = [a_{ij}]$ is a $(m \times n)$ matrix, $b = (b_1, b_2, \ldots, b_m)$ is a column vector, and 0 is an n-dimensional null column vector. In other words, the problem is to

$$\text{minimize } cX \tag{10.2}$$

subject to

$$X \geqslant 0, \tag{10.3}$$

$$AX = b. \tag{10.4}$$

A *feasible solution* to the linear programming problem is a vector $X = (x_1, x_2, \ldots, x_n)$ that satisfies conditions (10.3) and (10.4).

A *basic solution* that satisfies condition (10.4) is obtained by putting $(n - m)$ variables equal to zero and solving for the remaining m variables, provided the equations are solvable. The m variables are called *basic variables*.

A *basic-feasible solution* is a basic solution that also satisfies condition (10.3).

A *non-degenerate basic-feasible solution* is a basic-feasible solution with exactly m positive x_i's; i.e., all the basic solutions are positive.

With the foregoing definitions, we shall now prove three theorems which describe the properties of a solution to the linear programming problem.

Theorem 1 The set of all feasible solutions to the linear programming problem is a convex set.

If the set of solutions has only one element, then the theorem is obviously true. Assume there are at least two solutions, U and V. We have to show that every convex combination of these two feasible solutions is also a feasible solution. We have

$$AU = b \quad \text{for } U \geqslant 0,$$

$$AV = b \quad \text{for } V \geqslant 0.$$

Let X be a convex combination of U and V so that

$$X = aU + (1 - a)V \quad \text{for } 0 \leqslant a \leqslant 1.$$

We observe that $X \geqslant 0$. To see whether it satisfies condition (10.4), we substitute and get

$$\begin{aligned} AX &= AaU + A(1 - a)V \\ &= aAU - aAV + AV \\ &= b. \end{aligned}$$

Hence, the theorem is proved.

Let the convex set of solutions to the linear programming problem be denoted by K. The region occupied by K can be void, or a convex polyhedron, or unbounded in some region. If K is void, the given problem does not have a solution. If K is a polyhedron, the problem has a solution which is finite. If K is unbounded, then the minimum might be unbounded. Generally, in practical linear programming problems, K is either a convex polyhedron or unbounded. We shall assume that K is a convex polyhedron. We have already observed that every point in a convex polyhedron can be expressed as a convex combination of its extreme points. In other words, we have the following corollary.

Corollary 1 Every feasible solution in K, which is a convex polyhedron, can be expressed as a convex combination of the extreme feasible solutions in K.

Theorem 2 The objective function (10.2) assumes its minimum at an extreme point of the convex polyhedron K, generated by the set of feasible solutions to the problem.

Let $E_1, E_2, \ldots, E_n$ be the extreme points of the convex polyhedron. If the objective function has a minimum at one of these extreme points, then the theorem is true. Otherwise, t X be a point inside the polyhedron where the objective function assumes a minimum. We can express X as a convex combination of the extreme points of K, i.e.,

$$X = a_1E_1 + a_2E_2 + \ldots + a_nE_n,$$

where a_i's $\geqslant 0$ and $a_1 + a_2 + \ldots + a_n = 1$.

Substituting in the objective function, which is linear, we get

$$\begin{aligned} cX &= c(a_1E_1 + a_2E_2 + \ldots + a_nE_n) \\ &= a_1cE_1 + a_2cE_2 + \ldots + a_ncE_n. \end{aligned}$$

Let the minimum value of the objective function (which occurs at X) be ρ. The values assumed by the objective function at the extreme points are $cE_1, cE_2, \ldots, cE_n$, and we have assumed that each one of these is greater than cX. Let the minimum among these be cE_m, occurring at the extreme point m. Then we have

$$a_1cE_1 + a_2cE_2 + \ldots + a_ncE_n \geqslant a_1cE_m + a_2cE_m + \ldots + a_ncE_m,$$

i.e.,

$$cX \geqslant cE_m(a_1 + a_2 + \ldots + a_n)$$

or

$$cX \geqslant cE_m = \rho.$$

This contradicts our assumption. Hence, there is an extreme point where the objective function assumes its minimum value.

If the objective function assumes a minimum at more than one extreme point, we get the following corollary.

Corollary 2 If the objective function assumes a minimum at more than one extreme point, then it takes on the same value for every convex combination of those particular points.

To prove this, let the objective function attain its minimum value at the extreme points $E_1, E_2, \ldots, E_p$. We have

$$cE_1 = cE_2 = \ldots = cE_p = \rho.$$

Let X be a convex combination of these points so that

$$X = a_1E_1 + a_2E_2 + \ldots + a_pE_p,$$

where all a_i's $\geqslant 0$ and $a_1 + a_2 + \ldots + a_p = 1$. Then

$$\begin{aligned} cX &= c(a_1E_1 + a_2E_2 + \ldots + a_pE_p) \\ &= a_1cE_1 + a_2cE_2 + \ldots + a_pcE_p \\ &= \rho(a_1 + a_2 + \ldots + a_p) \\ &= \rho. \end{aligned}$$

Hence, the corollary is proved.

Therefore, by Theorem 2, we consider only the extreme points of K in our search for a minimum of the linear programming problem. We shall now show that a basic-feasible solution corresponds to an extreme point solution.

Theorem 3 If we assume that the linear programming problem as stated in constraints (10.3) and (10.4) is solvable, a basic-feasible solution to the problem corresponds to an extreme point of the convex set of feasible solutions.

Let $X = (x_1, x_2, \ldots, x_m, 0, \ldots, 0)$ be a basic-feasible solution. Suppose X is not an extreme point. Then it can be written as a convex combination of two other points, X' and X'' in K, such that $X = \alpha X' + (1 - \alpha)X''$ for $0 < \alpha < 1$. Since all the elements of X are non-negative, and $0 < \alpha < 1$, the last $(n - m)$ elements of X' and X'' must also equal zero, i.e.,

$$X' = (x_1', x_2', \ldots, x_m', 0, \ldots, 0),$$
$$X'' = (x_1'', x_2'', \ldots, x_m'', 0, \ldots, 0).$$

Since X' and X'' are feasible solutions, we have

$$AX' = b, \qquad AX'' = b.$$

Subtracting one from the other and expanding, we get

$$a_{11}(x_1' - x_1'') + a_{12}(x_2' - x_2'') + \ldots + a_{1m}(x_m' - x_m'') = 0$$
$$a_{21}(x_1' - x_1'') + a_{22}(x_2' - x_2'') + \ldots + a_{2m}(x_m' - x_m'') = 0$$
$$\vdots$$
$$a_{m1}(x_1' - x_1'') + a_{m2}(x_2' - x_2'') + \ldots + a_{mm}(x_m' - x_m'') = 0$$

These are a set of m homogeneous equations in m unknowns. Hence,

$$x_1' = x_1'', x_2' = x_2'', \ldots, x_m' = x_m''.$$

This means that X cannot be expressed as a convex combination of two other distinct points in K and hence it must be an extreme point of K.

We now conclude that

(i) there is an extreme point of K at which the objective function assumes its minimum value; and

(ii) every basic-feasible solution corresponds to an extreme point of K.

These two conclusions support the fundamental theorem proved in Section 4.6. According to the fundamental theorem, if the linear programming problem stated by the objective function (10.2) and conditions (10.3) and (10.4) can be solved, it will have an optimal solution in which, *at most*, m of the admissible variables are positive.

The reader is referred to the proof given in Section 4.6. The Simplex criterion described in Section 4.7 helps us to move from one extreme point solution to the next better extreme point solution until the optimal solution is determined.

10.6 DUALITY THEOREM

A linear programming problem with m inequalities in n structural variables,

$$AX \geqslant b,$$
$$X \geqslant 0, \tag{10.5}$$
$$C^*X = Z \text{ (a minimum)},$$

where A is the $(m \times n)$ matrix of coefficients a_{ij}, X is the column vector x_j, b is the column vector b_i, and C^* is the row vector c_j, has a dual with n inequalities in n structural variables,

$$\begin{aligned} &A^*Y \leqslant C, \\ &Y \geqslant 0, \\ &b^*Y = w(\text{a maximum}), \end{aligned} \tag{10.6}$$

where A^* is the transpose of A, Y is a column vector of variables $y_1, y_2, \ldots, y_m$, and b^* is the transpose of b. Changing the inequalities in sets (10.5) and (10.6) into equalities by using the slack variables $x_1', x_2', \ldots, x_m'$ and $y_1', y_2', \ldots, y_n'$, respectively, we get

$$\begin{aligned} &AX - I_mX' = b, \\ &X \geqslant 0, \\ &X' \geqslant 0, \\ &C^*X = Z(\text{a minimum}), \end{aligned} \tag{10.7}$$

$$\begin{aligned} &A^*Y + I_nY' = C, \\ &Y \geqslant 0, \\ &Y' \geqslant 0, \\ &b^*Y = w(\text{a maximum}), \end{aligned} \tag{10.8}$$

where I_m, I_n are unit matrices, X', Y' are column vectors of the slack variables $x_1', x_2', \ldots, x_m'$ and $y_1', y_2', \ldots, y_n'$, respectively.

The duality theorem states that in the optimal basic solutions of sets (10.7) and (10.8)

(i) the values of x_j and x_i' in the basis are numerically equal to the Simplex coefficients of y_j' and y_i; and the values of y_i and y_j' in the basis are numerically equal to the Simplex coefficients of x_i' and x_j, respectively; and

(ii) $Z_{\min} = w_{\max}$.

Proof We can distinguish two cases. In one case, the optimal basis for the primal consists of only the structural (or decision) variables and no slack variables enter into it. In the second, the optimal basis for the primal contains some slack variables in addition to the structural variables.

We shall consider these two cases separately.

Case (1) Let the optimal basis consist of only the structural variables. Partition the vector X into a vector X_1 of m basic variables and a vector X_2 of the remaining $(n - m)$ structural variables. Let A and C^* be partitioned correspondingly. Then set (10.7) can be written as

$$\begin{aligned} &A_1X_1 + A_2X_2 - I_mX' = b, \\ &X_1 \geqslant 0, \qquad X_2 \geqslant 0, \qquad X' \geqslant 0, \\ &C_1^*X_1 + C_2^*X_2 = Z \text{ (a minimum)}. \end{aligned} \tag{10.9}$$

In set (10.9), A_1 is an $(m \times m)$ matrix, A_2 is an $m \times (n - m)$ matrix, and X_1, X_2 are column vectors having m and $(n - m)$ elements, respectively. Solving the set for X_1, we get

$$X_1 = A_1^{-1}b - A_1^{-1}A_2X_2 + A_1^{-1}X' \tag{10.10}$$

and, substituting in the expression for Z, we obtain

$$Z = C_1^*A_1^{-1}b - (C_1^*A_1^{-1}A_2 - C_2^*)X_2 + C_1^*A_1^{-1}X'. \tag{10.11}$$

Since we have assumed that X_1 is the optimal basis, Eqs. (10.10) and (10.11) give the optimal solutions by putting $X_2 = 0$ and $X' = 0$. For Z to be a minimum, the coefficient $(C_1^*A_1^{-1}A_2 - C_2^*)$ is negative and the coefficient $C_1^*A_1^{-1}$ is positive.

We can rewrite set (10.9) similarly after partitioning I_n and Y' as

$$\begin{aligned} &A_1^*Y + I_mY_1' = C_1, \\ &A_2^*Y + I_{n-m}Y_2' = C_2, \\ &Y \geqslant 0, \qquad Y_1' \geqslant 0, \qquad Y_2' \geqslant 0, \\ &b^*Y = w \text{ (a maximum)}. \end{aligned} \tag{10.12}$$

In set (10.12), A_1^*, the transpose of A_1, is an $(m \times m)$ matrix, A_2^* is an $(n - m) \times m$ matrix, Y is a column vector with m elements, and Y_1', Y_2' are column vectors with m and $(n - m)$ elements, respectively. C_1 and C_2 are column vectors with m and $(n - m)$ elements, respectively. Solving for Y and Y_2' in terms of Y_1', we get

$$\begin{aligned} &Y = (A_1^*)^{-1}C_1 - (A_1^*)^{-1}Y_1', \\ &Y_2' = [-A_2^*(A_1^*)^{-1}C_1 + C_2] + A_2^*(A_1^*)^{-1}Y_1'. \end{aligned} \tag{10.13}$$

Substituting for w, we have

$$w = b^*(A_1^*)^{-1}C_1 - b^*(A_1^*)^{-1}Y_1'. \tag{10.14}$$

From Eqs. (10.10), (10.11), (10.13), and (10.14), we observe:

(i) The values of X_1, X_2, and X' in the optimal basis of the primal are

$$X_1 = A_1^{-1}b, \qquad X_2 = 0, \qquad X' = 0,$$

and these are numerically equal to the (transposed) Simplex coefficients of Y_1', Y_2', and Y in the dual.

(ii) The values of Y, Y_1', and Y_2' in the optimal basis of the dual are

$$Y = (A_1^*)^{-1}C_1, \qquad Y_1' = 0, \qquad Y_2' = -A_2^*(A_1^*)^{-1}C_1 + C_2,$$

and these are numerically equal to the (transposed) Simplex coefficients of X', X_1, and X_2 in the dual.

Further, the Simplex coefficient $b^*(A_1^*)^{-1}$ in the dual is positive, since this is the value of X_1 in the basis, set (10.10). Hence, in the optimal solution,

$$Z_{\min} = w_{\max}.$$

Case (2) The optimal basis for the dual contains some slack variables in addition to the structural variables. In this case, we shall be making use of the complementary slackness theorem discussed in Section 6.3. A part of the theorem states:

If in the optimal solution of a primal, a slack variable x_j' has a non-zero value, then the j-th variable of its dual has a zero value in the optimal solution of the dual.

In set (10.7), there are $(n + m)$ admissible variables of which m are slack variables. We shall rewrite set (10.7) as

$$\begin{aligned} &\bar{A}\bar{X} = b, \\ &\bar{X} \geqslant 0, \qquad\qquad (10.15) \\ &\bar{C}^*X = Z \text{ (a minimum).} \end{aligned}$$

Here $\bar{A}$ is an $m \times (n + m)$ matrix and $\bar{X}$ is a column vector consisting of $(n + m)$ elements. In this, we have included the slack variables also and we have not identified them by primes. $\bar{C}^*$ is a row vector with $(n + m)$ elements of which the coefficients corresponding to the slack variables are zero. Let there be r slack variables appearing among the m basic variables in the optimal solution of the problem. On the same lines as in Case (1), we can rewrite set (10.15) as

$$\begin{aligned} &\bar{A}_1\bar{X}_1 + \bar{A}_2\bar{X}_2 = b, \\ &\bar{X}_1 = 0, \qquad \bar{X}_2 = 0, \qquad\qquad (10.16) \\ &\bar{C}_1^*\bar{X}_1 + \bar{C}_2^*\bar{X}_2 = Z \text{ (a minimum).} \end{aligned}$$

In set (10.16), $\bar{X}_1$ corresponds to the basis and is a column vector with m elements (including r slack variables), $\bar{A}_1$ is an $(m \times m)$ matrix, $\bar{A}_2$ is an $(m \times n)$ matrix, and $\bar{X}_2$ is a column vector containing n elements. Solving for $\bar{X}_1$ in terms of $\bar{X}_2$, we get

$$\bar{X}_1 = \bar{A}_1^{-1}b - \bar{A}_1^{-1}\bar{A}_2\bar{X}_2 \qquad (10.17)$$

and, substituting in the expression for Z, we have

$$Z = \bar{C}_1^*\bar{A}_1^{-1}b - (\bar{C}_1^*\bar{A}_1^{-1}\bar{A}_2 - \bar{C}_2^*)\bar{X}_2. \qquad (10.18)$$

We get the optimal solutions by putting $\bar{X}_2 = 0$. If we form the dual to set (10.16), we get

$$\begin{aligned}
&\bar{A}_1^* Y \leqslant \bar{C}_1, \\
&\bar{A}_2^* Y \leqslant \bar{C}_2, \\
&Y \text{ is unrestricted in sign,} \qquad\qquad (10.19) \\
&b^* Y = w \text{ (a maximum).}
\end{aligned}$$

Introducing slack variables, we get

$$\begin{aligned}
&\bar{A}_1^* Y + I_m Y_1' = \bar{C}_1, \\
&\bar{A}_2^* Y + I_n Y_2' = \bar{C}_2, \qquad\qquad (10.20) \\
&\bar{Y}_1' \geqslant 0, \qquad Y_2' \geqslant 0.
\end{aligned}$$

Here Y is a column vector with m elements. Among these, r elements are zero according to the complementary slackness theorem. Further, $\bar{C}_1$ is also a column vector with m elements, r of them being equal to zero. This means that the corresponding r elements in Y_1' will be equal to zero.

From set (10.20), solving for Y and Y_2' in terms of Y_1', we get

$$\begin{aligned}
&Y = (\bar{A}_1^*)^{-1} C_1 - (A_1^*)^{-1} Y_1', \\
&Y_2' = -\bar{A}_2^* (\bar{A}_1^*)^{-1} C_1 + C_2 + A_2^* (A_1^*)^{-1} Y_1'.
\end{aligned}$$

Substituting for w, we get

$$w = b^* (\bar{A}_1^*)^{-1} C_1 = b^* (\bar{A}_1^*)^{-1} Y_1'.$$

It is obvious that for Case (2) also the duality theorem is satisfied.

11

Introduction to Theory of Games

11.1 INTRODUCTION

The optimization problems considered so far involved only one objective function which needed either maximization or minimization. These situations correspond to problems involving one decision-maker who is interested in either maximizing his profit or minimizing his loss. However, in actual life a large number of situations involve two or more decision-makers who have conflicting interests. For example, if three manufacturers compete in the production of a similar product, then it is easy to visualize a situation where each one plans to surpass his rivals. While most business practices in the competitive field are characterized by such conflicts, the analysis of real situations becomes extremely complicated. However, an insight into such problems can be obtained by considering a simpler situation involving only two decision-makers. If the gain of one decision-maker is the loss of the other, then the problem is generally classified as a *two-person zero-sum game.*

In a two-person zero-sum game, the interests of the two players (decision-makers) are totally conflicting. Consequently, these games reflect war situations. The theory of games of strategy was first proposed by the French mathematician Emile Borel in 1921. The solution to the problem, in the form of a minimax theorem, was given by Von Neumann in 1928. The Game Theory as a method for analyzing competitive situations in economics, warfare, and other areas of conflicting interests was developed by Von Neumann and Oskar Morganstern. The Simplex method of Dantzig provides an elegant analysis of such problems. In this chapter, we shall deal with two-person zero-sum games, which are also called *Matrix Games.*

11.2 2×2 MATRIX GAMES

Let us consider two players, A and B, in a game of coin-tossing. The result for A, and so too for B, can be either heads or tails. The rules of the game are as follows.

If the outcome for A coincides with that for B (i.e., heads for both or tails for both), then A wins a rupee from B. On the other hand, if the outcome for A does not match that for B (i.e., heads for A and tails for B or vice versa), then B wins a rupee from A. The pay-off for A can be represented in the form of a matrix as in Table 11.1. This pay-off matrix

TABLE 11.1

		Player B	
		H	T
Player A	H	1	−1
	T	−1	1

represents the amount paid to A by B for each of the four possible outcomes (H, H), (H, T), (T, H), and (T, T).

Now, let us consider the case when the game is played 100 times. If A gets heads 60 times in 100 tosses, then the frequency of heads turning up is said to be $60/100 = 0.6$; the frequency of tails turning up is then $40/100 = 0.4$. When the number of tosses is large, the frequency is referred to in terms of probability. In the tossing of a fair coin (i.e., a perfect coin tossed without bias), heads and tails will turn up with equal frequency in the long run; in other words, the probability of getting either heads or tails is 0.5. For a long-run game, the matrix appears as shown in Table 11.2. Here the probability (frequency) of heads or tails being turned up by players A and B is also indicated.

TABLE 11.2

			Player B	
			H	T
			1/2	1/2
Player A	H	1/2	1	−1
	T	1/2	−1	1

When the game is played many times, A plays heads on half the number of occasions and tails on the remaining half. Since B is also playing independently with equal frequency, A wins half the number of games and loses on the remaining occasions. A's *average winnings per play* can be calculated as follows.

Let us assume that the game is played N times, where N is large. On $N/2$ occasions, A turns up heads. Since B is playing the game independently, on these $N/2$ occasions, B plays heads $N/4$ times and tails $N/4$ times. Thus, the winnings of A on $N/2$ occasions are

$$\frac{N}{4}(+1) + \frac{N}{4}(-1) = \frac{N}{2}[\tfrac{1}{2}(+1) + \tfrac{1}{2}(-1)].$$

Similarly, on the remaining $N/2$ occasions, A plays tails and B once again turns up heads $N/4$ times and tails $N/4$ times. A's winnings are now

$$\frac{N}{4}(-1) + \frac{N}{4}(+1) = \frac{N}{2}[\tfrac{1}{2}(-1) + \tfrac{1}{2}(1)].$$

The average winnings of A per play are therefore

$$\frac{1}{N}\{\frac{N}{2}[\tfrac{1}{2}(+1) + \tfrac{1}{2}(-1)] + \frac{N}{2}[\tfrac{1}{2}(-1) + \tfrac{1}{2}(+1)]\}$$
$$= \tfrac{1}{2}[\tfrac{1}{2}(+1) + \tfrac{1}{2}(-1)] + \tfrac{1}{2}[\tfrac{1}{2}(-1) + \tfrac{1}{2}(+1)] = 0.$$

Let A now change his strategy: he plays heads N_1 times and tails N_2 times so that $N_1 + N_2 = N$. Let B continue turning up heads and tails with equal frequency. The winnings of A are then given by

$$[\tfrac{1}{2}N_1(+1) + \tfrac{1}{2}N_1(-1)] + [\tfrac{1}{2}N_2(-1) + \tfrac{1}{2}N_2(+1)]$$
$$= N_1[\tfrac{1}{2}(+1) + \tfrac{1}{2}(-1)] + N_2[\tfrac{1}{2}(-1) + \tfrac{1}{2}(+1)].$$

The average winnings of A per play are now given by

$$\frac{N_1}{N}[\tfrac{1}{2}(+1) + \tfrac{1}{2}(-1)] + \frac{N_2}{N}[\tfrac{1}{2}(-1) + \tfrac{1}{2}(+1)]$$
$$= p[\tfrac{1}{2}(+1) + \tfrac{1}{2}(-1)] + q[\tfrac{1}{2}(-1) + \tfrac{1}{2}(+1)] = 0.$$

When N is large, the ratio $N_1/N = p$ is the *probability* of heads being played by A. Similarly, q is the probability of tails being played by A. We note that $p + q = 1$.

It is important to observe that whatever the strategy of A, his *long-run expectation* will be zero as long as B adopts an equal frequency strategy (1/2, 1/2). Similarly, A, by following the equal frequency strategy, can ensure that the long-run expectation of B will be zero.

Suppose, now, A changes his strategy from (1/2, 1/2) to (2/3, 1/3) and B adopts the strategy (1/4, 3/4). Let the pay-off matrix for A remain the same as before. A's average winnings per play are then expressed as

$$\tfrac{2}{3}[\tfrac{1}{4}(+1) + \tfrac{3}{4}(-1)] + \tfrac{1}{3}[\tfrac{1}{4}(-1) + \tfrac{3}{4}(+1)] = (\tfrac{2}{3})(-\tfrac{1}{2}) + (\tfrac{1}{3})(\tfrac{1}{2}) = -\tfrac{1}{6}.$$

Hence, A loses to B at an average rate of Rs $\frac{1}{6}$ per play. Of course, the same result can be obtained by calculating B's average winnings per play:

$$\tfrac{1}{4}[\tfrac{2}{3}(-1) + \tfrac{1}{3}(+1)] + \tfrac{3}{4}[\tfrac{2}{3}(+1) + \tfrac{1}{3}(-1)] = (\tfrac{1}{4})(-\tfrac{1}{3}) + (\tfrac{3}{4})(\tfrac{1}{3}) = \tfrac{1}{6}.$$

This is a gain for B and a loss for A.

In any game, both A and B tend to be clever and each adopts a strategy to excel the other. In such a situation, it can be shown that (1/2, 1/2) is indeed the optimal strategy for both players. To see this, let the strategy of A be $(x, 1 - x)$ and that of B be $(y, 1 - y)$. It should be observed that, according to the rules of probability, the sum of the factors appearing in each strategy should be equal to one, i.e., $x + (1 - x) = 1$ and $y + (1 - y) = 1$. Let the pay-off matrix for A be as in Table 11.3, i.e., the same as in

TABLE 11.3

			Player B	
			H	T
			y	$(1-y)$
Player A	H	x	1	−1
	T	$(1-x)$	−1	1

Table 11.1. Let Z be the average winnings of A per play. This value is given by

$$Z = x[y(+1) + (1 - y)(-1)] + (1 - x)[y(-1) + (1 - y)(+1)]$$

$$= 4xy - 2x - 2y + 1$$

which can be written as

$$Z = (2x - 1)(2y - 1) = 4(x - \tfrac{1}{2})(y - \tfrac{1}{2}). \tag{11.1}$$

The desire of A is to maximize his expected gain Z, whereas B aims to maximize *his* expected gain which is $-Z$, i.e., minimize Z.

Equation (11.1) shows that $Z = 0$ when either $x = 1/2$ or $y = 1/2$, indicating that when either A or B adopts an equal frequency strategy, there is no net loss or gain for either of them. Let us consider the case when A chooses a strategy where $x > 1/2$. Then $(x - 1/2)$ will be positive. But B can make Z negative by choosing a strategy that makes $(y - 1/2)$ negative. The most negative value is achieved when $y = 0$. Thus, when A opts for a strategy where $x > 1/2$, B's most rewarding strategy will be (0, 1). A's expected gain is then $-(2x - 1)$. On the other hand, if A chooses a strategy where $x < 1/2$, then B's best strategy will be (1, 0). The particular strategies (1, 0) and (0, 1) are called *pure strategies*. As per these strategies, the player opts to turn up only heads or only tails. A strategy of the type (1/4, 3/4), (3/8, 5/8), . . . is called a *mixed strategy*.

11.2.1 Graphical Representation of Eq. (11.1)

Equation (11.1) can be represented in a graphical form as shown in Fig. 11.1. The figure is drawn in a three-dimensional format. The surface is known as a hyperbolic paraboloid or a *saddle surface*. O′ is the *saddle point* having the coordinates (1/2, 1/2, 0). Curve O′A is a parabola; this lies below the xy-plane and its highest point is O′. Curve O′B is also a parabola; it curves upward above the xy-plane and its lowest point is O′. The value of Z at O′, which is zero, is the lowest value for the curve O′B and, at the same time, the highest value for another curve O′A. Hence, O′ is called the *minimax point* of the function Z.

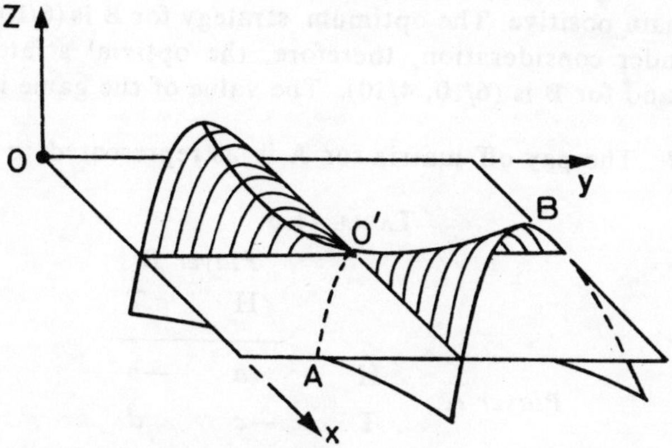

FIGURE 11.1

So far, we have considered a pay-off matrix with unit values. This is not always so. Very often, the loss of a particular game may be more severe to one player than to the other. For example, if there are two warring nations, the loss of a dozen striker aircraft may affect one nation much more severely than the other. Consider the pay-off matrix shown in

TABLE 11.4

			Player B	
			H	T
			u	$(1-u)$
Player A	H	x	1	−2
	T	$(1-x)$	−3	4

Table 11.4. Here A's expected gain per play will be

$$\begin{aligned} Z &= x[u(+1) + (1-u)(-2)] + (1-x)[u(-3) + (1-u)(4)] \\ &= 10xu - 6x - 7u + 4 \\ &= 10(xu - \tfrac{6}{10}x - \tfrac{7}{10}u) + 4 \\ &= 10(x - \tfrac{7}{10})(u - \tfrac{6}{10}) - 0.2. \end{aligned}$$

This equation reveals that if B chooses the strategy $u = 6/10$, he will always gain Rs 0.2 per game regardless of the type of strategy A adopts. A can merely ensure that his loss will be a minimum by adopting the strategy (7/10, 3/10). Because if A chooses to play with $x > 7/10$, then B can choose $u < 6/10$, forcing A to lose more than Rs 0.2 per game. On the other hand, if A chooses $x < 7/10$, then B can opt for $u > 6/10$, once again forcing A to lose. Hence, for A, the optimal strategy is (7/10, 3/10). Similarly, if B opts to play with $u > 6/10$, A can choose $x < 7/10$, thus making his gain positive. The optimum strategy for B is (6/10, 4/10). For the game under consideration, therefore, the optimal strategy for A is (7/10, 3/10) and for B is (6/10, 4/10). The value of the game is Rs -0.2.

Problem 11.1 The pay-off matrix for A is as represented in Table 11.5.

TABLE 11.5

		Player B	
		H	T
Player A	H	a	$-b$
	T	$-c$	d

Let the strategy for A be (x, y) and that for B be (u, v), where $x + y = u + v = 1$, x, y, u, and v each is $\geqslant 0$. The expected gain per play for A is

$$\begin{aligned} Z &= x(ua - vb) + y(-uc + vd) \\ &= x[ua - (1 - u)b] + (1 - x)[-uc + (1 - u)d] \\ &= xu(a + b + c + d) + x(-b - d) + u(-c - d) + d \\ &= (a + b + c + d)(x - \alpha)(u - \beta) + d - \frac{\alpha\beta}{a + b + c + d}, \end{aligned}$$

where

$$\beta = \frac{b + d}{a + b + c + d}, \qquad \alpha = \frac{c + d}{a + b + c + d}.$$

Therefore, the optimal strategy for A is $(\alpha, 1 - \alpha)$ and for B is $(\beta, 1 - \beta)$. The value of the game is

$$V = d - \frac{\alpha\beta}{a + b + c + d} = \frac{ad - bc}{a + b + c + d}.$$

11.3 OPTIMAL PURE STRATEGIES

Let us consider the pay-off matrix shown in Table 11.6. Let (p, q) be the

TABLE 11.6

			Player B	
			H	T
Player A	H	p	a	b
	T	q	c	d

strategy of A, with $p + q = 1$. If player B adopts the pure strategy (1, 0), the pay-off for A is

$$x = p[(1(a) + 0(b)] + q[1(c) + 0(d)]$$

or

$$x = pa + qc. \tag{11.2}$$

On the other hand, if B adopts the pure strategy (0, 1), the pay-off for A is

$$y = p[0(a) + 1(b)] + q[0(c) + 1(d)]$$

or

$$y = pb + qd. \tag{11.3}$$

x and y are the pay-offs for A corresponding to the pure strategies (1, 0) and (0, 1) adopted by B. Now, let B opt for the strategy (u, v), with $u + v = 1$. Accordingly, the pay-off for A becomes

$$\begin{aligned} Z &= p(ua + vb) + q(uc + qd) \\ &= u(pa + qc) + v(pb + qd), \end{aligned}$$

i.e.,

$$Z = ux + vy. \tag{11.4}$$

When A adopts a strategy (p, q), the objective of B will be to adopt that strategy which minimizes the pay-off to A. It can be shown that this minimum pay-off will always be the smaller of the two pay-offs x and y. It should be remembered that x and y are the pay-offs to A when B opts for one or the other of the two pure strategies (1, 0) and (0, 1). To see this, let $x \leqslant y$. Then

$$x = (u + v)x = ux + vx \leqslant ux + vy \leqslant uy + vy = y.$$

Thus, when $x \leqslant y$,

$$x \leqslant ux + vy \leqslant y. \tag{11.5}$$

Similarly, when $y \leqslant x$,

$$y \leqslant ux + vy \leqslant x. \tag{11.6}$$

Therefore, for all possible strategies (u, v) of B, the minimum pay-off to A will be that corresponding to the smaller of the two pay-offs x and y, which B would have paid had he opted for one of the two pure strategies.

11.4 $m \times n$ MATRIX GAME

So far, we have dealt with 2×2 matrix games, where A and B each had two strategies, (p, q) and (u, v) with $p + q = u + v = 1$. It is necessary to recall that p, q, u, and v are the probabilities (or relative frequencies) of A's and B's moves. The tossing of a coin conveniently fitted into this structure. Obviously, this game can be extended to accommodate m possible moves by A and n possible moves by B. The pay-off matrix for A can then be represented by an $m \times n$ matrix:

$$\begin{array}{c} \text{B-column number} \\ \begin{array}{ccc} & & \begin{array}{cccc} 1 & 2 & \dots & n \end{array} \\ \text{A-row number} & \begin{array}{c} 1 \\ 2 \\ \vdots \\ m \end{array} & \begin{bmatrix} a_{11} & a_{12} & \dots & a_{1n} \\ a_{21} & a_{22} & \dots & a_{2n} \\ \vdots & \vdots & & \vdots \\ a_{m1} & a_{m2} & \dots & a_{mn} \end{bmatrix} = [a_{ij}]. \end{array} \end{array} \tag{11.7}$$

Here a_{ij} is the general element of the matrix. This represents the pay-off to A when A makes a move corresponding to the i-th row and B makes a move corresponding to the j-th column. In a total number of N games, let us say that A makes the following number of moves associated with each row:

move associated with 1st row	m_1
move associated with 2nd row	m_2
$\vdots$	$\vdots$
move associated with i-th row	m_i
$\vdots$	$\vdots$
move associated with m-th row	m_m

The relative frequency with which these moves are made is m_1/N, $m_2/N, \dots, m_m/N$. These moves are made in a random manner. When N becomes large, the relative frequency becomes the probability $p_1, p_2, \dots, p_m$. In other words, in the long run, p_1 is the probability of A making a move associated with the 1st row, p_2 is the probability of A making a move associated with the 2nd row, and so on. These probabilities are non-negative numbers:

$$p_1 + p_2 + \dots + p_m = 1. \tag{11.8}$$

The set of non-negative numbers $(p_1, p_2, \dots, p_m)$ is said to be the *mixed strategy* for player A. Similarly, for player B, the *mixed strategy* will be the set of numbers $(q_1, q_2, \dots, q_n)$, where the q_i's are non-negative:

$$q_1 + q_2 + \dots + q_n = 1. \tag{11.9}$$

If in the strategy $(p_1, p_2, \dots, p_m)$ all the numbers other than p_i are zero, and $p_i = 1$, it is known as the i-th pure strategy of A. In other words, the i-th pure strategy of A is $(0, 0, \dots, p_i = 1, \dots, 0)$. Similarly, the j-th pure strategy of B is

$$q_1 = 0, q_2 = 0, \ldots, q_j = 1, \ldots, q_n = 0$$

or

$$(0, 0, \ldots, q_j = 1, \ldots, 0).$$

11.4.1 Minimax Principle

Let us consider the following pay-off matrix for player A:

$$\begin{bmatrix} 2 & 10 & 0 & 8 \\ 4 & 2 & 6 & 6 \\ 8 & 4 & -2 & 0 \end{bmatrix} = [a_{ij}].$$

From the given matrix, we observe that

$$\min a_{1j} = a_{13} = 0, \qquad \min a_{2j} = a_{22} = 2, \qquad \min a_{3j} = a_{33} = -2.$$

Player A can of course select any strategy i, and for each i there is a minimum pay-off. Consequently, he can choose that pure strategy i for which the pay-off is the maximum among all the minimums. For example, if he makes a move corresponding to the first row, the minimum pay-off that he can expect is 0, regardless of the strategy B chooses. Similarly, if A selects a move corresponding to the second row, his minimum pay-off is 2, irrespective of the move selected by B. A's objective will obviously be to opt for that pure strategy i that offers the maximum among all the minimum pay-offs. From the given matrix,

$$\max_i \min_j a_{ij} = a_{22} = 2.$$

This is called the *lower value* of the game.

From the viewpoint of B, for each move j, there is a maximum pay-off to A. His objective of course is to select that pure strategy j for which the pay-off is the minimum among the possible maximum pay-offs. For the given 3×4 matrix,

$$\max a_{i1} = a_{31} = 8, \qquad \max a_{i2} = a_{12} = 10,$$

$$\max a_{i3} = a_{23} = 6, \qquad \max a_{i4} = a_{14} = 8.$$

Hence,

$$\min_j \max_i a_{ij} = a_{23} = 6.$$

This is the upper value of the game. If

$$\max_i \min_j a_{ij} = \min_j \max_i a_{ij} = w, \tag{11.10}$$

then A can be sure of winning w, and he can be prevented by B from winning more than w. In a matrix game where Eq. (11.10) holds good, each player can make sure that, by adopting a pure strategy, the winner gets the maximum among the minimums and the loser pays the minimum among the possible maximums. Such a pure strategy is called an *optimal pure strategy*. For the problem under discussion, the optimal pure strategy for A is (0, 1, 0), and for B is (0, 0, 1, 0). In this example, the values of $\max_i \min_j a_{ij}$ and $\min_j \max_i a_{ij}$ are quite different. However, for the matrix game

$$\begin{bmatrix} 6 & 10 & 18 \\ 2 & 4 & 6 \\ 0 & 14 & 8 \end{bmatrix} = [a_{ij}],$$

$$\max_i \min_j a_{ij} = \min_j \max_i a_{ij} = a_{11} = 6.$$

The element a_{11} is the minimum element in its row and is also the maximum element in its column. Such a point is called the *saddle point*. This aspect has already been illustrated in Fig. 11.1.

It is important to note one usual aspect of the maximin and minimax principles: we have observed that when A adopts the maximin strategy he can be sure that his gain will not be less than the maximin value (i.e., the lower value of the game). Hence, for any strategy of B,

$$\text{maximin value} \leqslant \text{pay-off to A}.$$

Similarly, when B adopts the minimax strategy, he can be sure that he will not lose more than the minimax value (i.e., the upper value of the game). Thus, for any strategy of A,

$$\text{minimax value} \leqslant \text{pay-off to A}.$$

Hence, combining the two, we have, for any strategy,

$$\text{maximin value} \leqslant \text{pay-off to A} = \text{minimax value of A and B}.$$

11.5 LONG-RUN EXPECTATION

Let us consider the $m \times n$ matrix game as shown in Table 11.7 where the

TABLE 11.7

		Player B		
		q_1	$q_2 \ldots q_n$	
Player A	p_1	a_{11}	$a_{12} \ldots a_{1n}$	
	p_2	a_{21}	$a_{22} \ldots a_{2n}$	
	$\vdots$	$\vdots$	$\vdots \quad \vdots$	
	p_m	a_{m1}	$a_{m2} \ldots a_{mn}$	

pay-off elements are a_{ij} and the mixed strategies of the two players are A$(p_1, p_2, \ldots, p_m)$, B$(q_1, q_2, \ldots, q_n)$. The long-run pay-off to A is

$$Z = p_1(q_1a_{11} + q_2a_{12} + \ldots + q_na_{1n}) + p_2(q_1a_{21} + q_2a_{22} + \ldots + q_na_{2n}) + \ldots + p_i(q_1a_{i1} + q_2a_{i2} + \ldots + q_na_{in}) + \ldots + p_m(q_1a_{m1} + q_2a_{m2} + \ldots + q_na_{mn}).$$

The general term in this sequence is

$$p_i(q_1a_{i1} + q_2a_{i2} + \ldots + q_na_{in}) = p_i \sum_{j=1}^{n} q_ja_{ij}.$$

The sequence Z is the sum of such expressions as given above with $i = 1, 2, \ldots, m$. Hence,

$$Z = \sum_{i=1}^{m} \sum_{j=1}^{n} p_ia_{ij}q_j. \tag{11.11}$$

This is also known as the *mathematical expectation* of A for the mixed strategies $(p_1, p_2, \ldots, p_m)$ and $(q_1, q_2, \ldots, q_n)$. It is written as

$$E(P, Q) = \sum_i \sum_j p_ia_{ij}q_j. \tag{11.12}$$

As an example, let us consider the pay-off matrix given as in Table 11.8 with the strategies A(0.2, 0.5, 0.3) and B(0.1, 0.4, 0.3, 0.2). Here the

TABLE 11.8

		Player B			
		0.1	0.4	0.3	0.2
	0.2	1	3	5	6
Player A	0.5	2	4	3	4
	0.3	4	3	3	5

mathematical expectation of A is

$$\begin{aligned} &0.2[0.1(1) + 0.4(3) + 0.3(5) + 0.2(6)] \\ &\quad + 0.5[0.1(2) + 0.4(4) + 0.3(3) + 0.2(4)] \\ &\quad + 0.3[0.1(4) + 0.4(3) + 0.3(3) + 0.2(5)] \\ &= 0.2(4.3) + 0.5(3.5) + 0.3(3.5) \\ &= 3.66. \end{aligned}$$

11.6 ADDITION OF CONSTANT TO $m \times n$ MATRIX ELEMENTS

Let us consider the pay-off matrix to A as shown with the strategies A$(p_1, p_2, \ldots, p_m)$ and B$(q_1, q_2, \ldots, q_n)$ in Table 11.9.

TABLE 11.9

		Player B	
		q_1	$q_2 \ldots q_n$
	p_1	a_{11}	$a_{12} \ldots a_{1n}$
Player A	p_2	a_{21}	$a_{22} \ldots a_{2n}$
	$\vdots$	$\vdots$	$\vdots \quad \vdots$
	p_m	a_{m1}	$a_{m2} \ldots a_{mn}$

By definition, the long-run winnings of A (i.e., the mathematical expectation) are given by Eq. (11.11) as

$$E_1(P, Q) = \sum_{i=1}^{m} \sum_{j=1}^{n} p_i q_j a_{ij}.$$

Let each element in the pay-off matrix be increased by a constant term w. Then a_{ij} becomes $(a_{ij} + w)$. The mathematical expectation is then

$$\begin{aligned} E_2(P, Q) &= \sum_{i=1}^{m} \sum_{j=1}^{n} p_i q_j (a_{ij} + w) \\ &= \sum_{i} \sum_{j} p_i q_j a_{ij} + w \sum_{i} \sum_{j} p_i q_j. \end{aligned}$$

The second term on the right-hand side is

$$\begin{aligned} w \sum_{i} \sum_{j} p_i q_j &= w \sum_{i} p_i (q_1 + q_2 + \ldots + q_n) \\ &= w(p_1 + p_2 + \ldots)(q_1 + q_2 + \ldots) \\ &= w \end{aligned}$$

since

$$\sum_{i=1}^{m} p_i = \sum_{j=1}^{n} q_j = 1.$$

Hence, the mathematical expectation for the modified pay-off matrix is

$$E_2(P, Q) = \sum_{i=1}^{m} \sum_{j=1}^{n} p_i q_j a_{ij} + w. \tag{11.13}$$

Hence, the optimal strategy for either A or B is not affected by the addition of a constant term w to the pay-off matrix.

The importance of the foregoing analysis is that the optimal strategies do not get affected if all the terms in the pay-off matrix can be made positive by the addition of a constant term. For example, the optimum strategies for the two pay-off matrices shown in Tables 11.10 and 11.11 are the same. This fact will be made use of in Chapter 12 to analyze a matrix game by means of linear programming techniques.

TABLE 11.10

		Player B			
		q_1	q_2	q_3	q_4
	p_1	1	−2	3	2
Player A	p_2	−3	1	−1	2
	p_3	0	2	−2	−3

TABLE 11.11

		Player B			
		q_1	q_2	q_3	q_4
	p_1	5	2	7	6
Player A	p_2	1	5	3	6
	p_3	4	6	2	1

EXERCISES

1. Let the strategy for A be (x, y) and for B be (u, v), where $x + y = 1$ and $u + v = 1$. For the pay-off matrices (a) and (b), determine the optimal strategy for both players and the value of each game.

(a)

		B	
		H	T
A	H	−1	2
	T	1	0

(b)

		B	
		H	T
A	H	−2	−4
	T	3	5

[*Ans.* (a) A: $(\frac{1}{4}, \frac{3}{4})$; B: $(\frac{1}{2}, \frac{1}{2})$; value: $\frac{1}{2}$; (b) A: (0, 1); B: (1, 0); value: 3.]

2. Let the pay-off matrix for A be as tabulated here. A adopts the strategy

		B	
		H	T
A	H	a	b
	T	c	d

(x, y) and B the strategy (u, v). As before, $x + y = u + v = 1$, and x, y, u, v, each $\geqslant 0$.

(a) Calculate the expected gain per play for A.

(b) Determine the effect on the expected gain when every one of the terms in the pay-off matrix is increased by k.

(c) What is the effect on the expected gain if the terms in the pay-off matrix are multiplied by k?

3. Show that, in an arbitrary pay-off matrix $[a_{ij}]$,

$$\max_i \min_j a_{ij} \leqslant \min_j \max_i a_{ij},$$

where

$$[a_{ij}] = \begin{bmatrix} a_{11} & a_{12} & \dots & a_{1n} \\ a_{21} & a_{22} & \dots & a_{2n} \\ \vdots & \vdots & & \vdots \\ a_{m1} & a_{m2} & \dots & a_{mn} \end{bmatrix}.$$

12

Linear Programming and Matrix Game

12.1 INTRODUCTION

In Chapter 11, we discussed the 2×2 matrix game in detail. While dealing with an $m \times n$ matrix game, the case of optimal pure strategy was mentioned and we concluded that such a matrix game can be optimally played by players A and B if

$$\max_i \min_j a_{ij} = \min_j \max_i a_{ij} = v.$$

However, not all matrix games can be optimally played by means of pure strategy. Generally, the optimal strategy from the point of view of A as well as B is a mixed strategy. In this chapter, we shall show the equivalence of a general $m \times n$ matrix game and the related linear programming problem.

12.2 PROBLEM STATEMENT

In a game played by A and B, A has m possible moves and B has n possible moves. The pay-off matrix for A is as shown in Table 12.1. The

TABLE 12.1

		Player B			
		q_1	q_2	$\cdots$	q_n
	p_1	a_{11}	a_{12}	$\cdots$	a_{1n}
Player A	p_2	a_{21}	a_{22}	$\cdots$	a_{2n}
	$\vdots$	$\vdots$	$\vdots$		$\vdots$
	p_m	a_{m1}	a_{m2}	$\cdots$	a_{mn}

probability that A makes the move corresponding to the i-th row is p_i and the probability that B makes the move corresponding to the j-th column is q_j. The objective of A is to adopt a strategy $(p_1, p_2, \ldots, p_m)$ such that his winnings are a maximum, whereas the objective of B is to adopt a strategy $(q_1, q_2, \ldots, q_n)$ such that his loss is a minimum. The problem therefore is to find the strategy of A and of B and also to determine the value of the game.

It will now be shown that in determining the best strategy for A it is sufficient for A to take into account only the pure strategies of his opponent B.

Let B adopt the pure strategy (1, 0, 0, . . . , 0). The long-run winnings (i.e., the mathematical expectation) of A are then

$$p_1a_{11} + p_2a_{21} + \ldots + p_ma_{m1} = \sum_{i=1}^{m} p_ia_{i1}.$$

If B adopts the pure strategy (0, 1, 0, . . . , 0), the long-run winnings of A are

$$p_1a_{12} + p_2a_{22} + \ldots + p_ma_{m2} = \sum_{i=1}^{m} p_ia_{i2}.$$

Similarly, for B's pure strategy $(0, 0, \ldots, q_j = 1, 0, \ldots, 0)$, the mathematical expectation of A is

$$p_1a_{1j} + p_2a_{2j} + \ldots + p_ma_{mj} = \sum_{i=1}^{m} p_ia_{ij}.$$

Instead of adopting a pure strategy, let B adopt a mixed strategy $(\bar{q}_1, \bar{q}_2, \ldots, \bar{q}_m)$ which he considers to be the best. Let the winnings for A under this condition be w. This means that if A adopts the strategy $(p_1, p_2, \ldots, p_m)$ which is the optimum against B's pure strategies, then the same optimum strategy should help A win at least w against B's mixed strategy, i.e.,

$$\begin{aligned}
p_1a_{11} + p_2a_{21} + \ldots + p_ma_{m1} &= \sum_i p_ia_{i1} \geqslant w \\
p_1a_{12} + p_2a_{22} + \ldots + p_ma_{m2} &= \sum_i p_ia_{i2} \geqslant w \\
&\vdots \\
p_1a_{1j} + p_2a_{2j} + \ldots + p_ma_{mj} &= \sum_i p_ia_{ij} \geqslant w \\
&\vdots \\
p_1a_{1n} + p_2a_{2n} + \ldots + p_ma_{mn} &= \sum_i p_ia_{in} \geqslant w
\end{aligned} \tag{12.1}$$

It can now be shown that if A adopts a strategy that fulfils conditions (12.1), he can be sure of winning at least w even if B adopts a mixed strategy. Under the mixed strategy $(q_1, q_2, \ldots, q_n)$ of B, the expectation of A is

$$\begin{aligned}
E(P, Q) &= \sum_i \sum_j p_iq_ja_{ij} \\
&= \sum_j q_j \sum_i p_ia_{ij} \\
&= q_1 \sum_i p_ia_{i1} + q_2 \sum_i p_ia_{i2} + \ldots + q_n \sum_i p_ia_{in}.
\end{aligned}$$

Using conditions (12.1), we get

$$E(P, Q) \geqslant q_1w + q_2w + \ldots + q_nw$$
$$= (q_1 + q_2 + \ldots + q_n)w,$$

i.e.,

$$E(P, Q) \geqslant w. \tag{12.2}$$

A's objective being to maximize his winnings, he opts for a strategy $(p_1, p_2, \ldots, p_m)$ which will ensure that his winnings will not at any rate be less than w, since B also tries to make sure that his losses will not be more than w. Conditions (12.1) and (12.2) show that if A optimizes his strategy against B's pure strategies, he can make sure that his winnings will not be less than w even if B adopts a mixed strategy. Hence, the problem for A is to determine $(p_1, p_2, \ldots, p_m)$ such that

$$p_1 + p_2 + \ldots + p_m = 1, \tag{12.3}$$

$$\left.\begin{aligned} &p_1a_{11} + p_2a_{21} + \ldots + p_ma_{m1} \geqslant w \\ &p_1a_{12} + p_2a_{22} + \ldots + p_ma_{m2} \geqslant w \\ &\vdots \\ &p_1a_{1n} + p_2a_{2n} + \ldots + p_ma_{mn} \geqslant w \end{aligned}\right\} \tag{12.4}$$

where w is the smallest expectation of A under B's best strategy, and w is to be maximized.

12.3 FORMULATION OF GAME PROBLEM AS LINEAR PROGRAMMING PROBLEM

We assume that every element appearing in the pay-off matrix is positive. If not, we can add a constant term to all the negative elements to make them positive. It has been observed in Section 11.6 that the addition of a constant term to the elements of the pay-off matrix does not alter the optimum strategy. Hence, w, the pay-off to A, can be taken to be greater than zero. Let us divide the terms on the left-hand and right-hand side of conditions (12.4) by w and let

$$p_1/w = x_1,\ p_2/w = x_2, \ldots, p_m/w = x_m. \tag{12.5}$$

It should be observed that

$$x_1 + x_2 + \ldots + x_m = \frac{1}{w}\sum_i p_i = \frac{1}{w}. \tag{12.6}$$

Also, since $p_i \geqslant 0$ and w is positive, $x_i \geqslant 0$. The problem of maximizing w is therefore equivalent to the minimization of $x_1 + x_2 + \ldots + x_m$. The problem for A (called the *primal problem*) can therefore be stated as:

$$\text{Minimize } Z = x_1 + x_2 + \ldots + x_m \tag{12.7}$$

subject to conditions (12.4) which in this case appear as

$$\begin{aligned} &a_{11}x_1 + a_{21}x_2 + \ldots + a_{m1}x_m \geqslant 1 \\ &a_{12}x_1 + a_{22}x_2 + \ldots + a_{m2}x_m \geqslant 1 \\ &\vdots \\ &a_{1n}x_1 + a_{2n}x_2 + \ldots + a_{mn}x_m \geqslant 1 \end{aligned} \tag{12.8}$$

$$x_1, x_2, \ldots, x_m \geqslant 0. \tag{12.9}$$

This is evidently a linear programming problem which can be solved by using the several techniques already described.

Considering B's problem, we find his objective is to choose a strategy that minimizes A's winnings. Whatever the strategy of A, B's strategy is not to lose more than w to A. If A adopts any of the pure strategies $(1, 0, \ldots, 0)$, $(0, 1, 0, \ldots, 0)$, $\ldots$, $(0, 0, \ldots, 1)$, then B has to ensure that his strategy $(q_1, q_2, \ldots, q_n)$ will not make him lose more than w to A, i.e.,

$$\begin{aligned} &q_1a_{11} + q_2a_{12} + \ldots + q_na_{1n} = \sum_j q_ja_{ij} \leqslant w \\ &q_1a_{21} + q_2a_{22} + \ldots + q_na_{2n} = \sum_j q_ja_{2j} \leqslant w \\ &\vdots \\ &q_1a_{m1} + q_2a_{m2} + \ldots + q_na_{mn} = \sum_j q_ja_{mj} \leqslant w \end{aligned} \tag{12.10}$$

It can again be shown that if conditions (12.10) are satisfied, then B can be sure that his loss will not exceed w even if A adopts a mixed strategy. To see this, let A's mixed strategy be $(p_1, p_2, \ldots, p_m)$. Then A's mathematical expectation (i.e., B's loss) is

$$\begin{aligned} E(P, Q) &= \sum_i \sum_j p_iq_ja_{ij} \\ &= \sum_i p_i \sum_j q_ja_{ij}. \end{aligned}$$

If conditions (12.10) are satisfied, then

$$E(P, Q) \leqslant \sum_i p_iw = w \sum_i p_i = w. \tag{12.11}$$

Hence, B's objective should be to optimize his strategy against A's pure strategies. Now, B's problem can be formulated as a linear programming problem. Let

$$q_1/w = y_1,\ q_2/w = y_2,\ \ldots,\ q_n/w = y_n. \tag{12.12}$$

Since $q_1, q_2, \ldots \geqslant 0$ and w is positive, $y_1, y_2, \ldots \geqslant 0$. From Eq. (12.12),

$$y_1 + y_2 + \ldots + y_n = \frac{1}{w} \sum_i q_i = \frac{1}{w}. \tag{12.13}$$

B's objective being to minimize w, this is equivalent to the maximization of $y_1 + y_2 + \ldots + y_n$. The problem of B (called the *dual problem*) can now be formulated as follows:

$$\text{Maximize } W = y_1 + y_2 + \ldots + y_n \tag{12.14}$$

subject to conditions (12.10) which in this case appear as

$$\begin{aligned} a_{11}y_1 + a_{12}y_2 + \ldots + a_{1n}y_n &\leqslant 1 \\ a_{21}y_1 + a_{22}y_2 + \ldots + a_{2n}y_n &\leqslant 1 \\ &\vdots \\ a_{m1}y_1 + a_{m2}y_2 + \ldots + a_{mn}y_n &\leqslant 1 \end{aligned} \tag{12.15}$$

$$y_1, y_2, \ldots, y_n \geqslant 0. \tag{12.16}$$

It is obvious that B's problem is actually the dual of A's problem, since it can be written down from Eq. (12.7) and conditions (12.8) and (12.9) by applying the usual rules of constructing a dual problem (see Chapter 6). A's problem involves n inequalities and m structural variables, whereas B's problem involves m inequalities and n structural variables. Whereas A's problem is one of minimization [equivalent to the maximization of his winnings as per Eq. (12.6)], B's problem is one of maximization of the objective function. As an éxample let us consider the following problem.

Problem 12.1 The pay-off matrix to A is as shown in Table 12.2. Determine the optimum strategy for A.

TABLE 12.2

		Player B	
		1	2
	1	2	5
Player A	2	3	1
	3	0	3

The problem is to determine the probabilities (i.e., the strategy) (p_1, p_2, p_3) and the value of w such that

$$\begin{aligned} &2p_1 + 3p_2 \geqslant w, \\ &5p_1 + p_2 + 3p_3 \geqslant w, \\ &p_1 + p_2 + p_3 \geqslant 1, \\ &p_1, p_2, p_3 = 0, \end{aligned}$$

and w is to be maximized. If we divide by w and put $p_1/w = x_1$, $p_2/w = x_2$, and $p_3/w = x_3$, the problem becomes:

Minimize $Z = x_1 + x_2 + x_3$

subject to

$$2x_1 + 3x_2 \geqslant 1,$$

$$5x_1 + x_2 + 3x_3 \geqslant 1,$$

$$x_1, x_2, x_3 \geqslant 0.$$

When we use slack variables, the constraints become

$$2x_1 + 3x_2 - x_4 = 1,$$

$$5x_1 + x_2 + 3x_3 - x_5 = 1,$$

$$x_1, x_2, x_3, x_4, x_5 \geqslant 0.$$

We can adopt x_4 and x_5 as the initial bases as in Tableau I and apply the Dual Simplex method to iterate the tableau.

TABLEAU I

Basis	x_1	x_2	x_3	x_4	x_5	
x_4	−2	−3		1		−1
x_5	−5	−1	−3		1	−1
	−1	−1	−1			0

To determine the new basis, we derive the ratios as in Table 1. Since

RATIO TABLE 1

non-basic variable	x_1	x_2	x_3
coefficient in last row	−1	−1	−1
coefficient in row 1	−2	−3	0
Ratio	1/2	1/3	

we have to choose the minimum ratio, x_2 replaces x_4 as the basis. The corresponding tableau appears as shown.

TABLEAU II

Basis	x_1	x_2	x_3	x_4	x_5	
x_2	2/3	1		−1/3		1/3
x_5	−13/3	0	−3	−1/3	1	−2/3
	−1/3	0	−1	−1/3		1/3

x_5 remains infeasible. We derive the ratios as in Table 2.

RATIO TABLE 2

non-basic variable	x_1	x_3	x_4
coefficient in last row	−1/3	−1	−1/3
coefficient in row 2	−13/3	−3	−1/3
Ratio	*1/13*	1/3	1

x_1 replaces x_5 as the basis. The corresponding tableau appears as now shown.

TABLEAU III

Basis	x_1	x_2	x_3	x_4	x_5	
x_2	0	1	−18/39	−15/39	6/39	3/13
x_1	1	0	9/13	1/13	−3/13	2/13
	0	0	−10/13	−12/13	−1/13	15/39

The coefficients in the last row of Tableau III indicate that the optimum solution has been reached and the feasible solutions are $x_1 = 2/13$, $x_2 = 3/13$, and $x_3 = 0$. Recalling that $p_1 = x_1 w$ and $p_2 = x_2 w$, we have

$$1 = p_1 + p_2 + p_3 = w(x_1 + x_2) = \tfrac{5}{13}w,$$

i.e.,

$$w = \tfrac{13}{5},$$

and $p_1 = 2/5$, $p_2 = 3/5$, $p_3 = 0$. Hence, the optimal strategy for A is (2/5, 3/5, 0), and the value of the game is 13/5.

It has been observed that B's problem is the dual of A's problem. Forming the dual to the primal, we have:

Maximize $W = y_1 + y_2$

subject to

$$2y_1 + 5y_2 \leqslant 1,$$

$$3y_1 + y_2 \leqslant 1,$$

$$3y_2 \leqslant 1,$$

$$y_1, y_2 \text{ each} \geqslant 0.$$

This is simpler to solve than the primal. Solving this graphically (see Fig. 12.1), we obtain $y_1 = 4/13$ and $y_2 = 1/13$. Recalling that $q_1 = wy_1$ and $q_2 = wy_2$, we get

$$q_1 = 4/5, \qquad q_2 = 1/5, \qquad w = 13/5.$$

Hence, B's optimal strategy is (4/5, 1/5) and he loses 13/5 to A. It should be observed that the lower value of the game (i.e., the maximin value) is

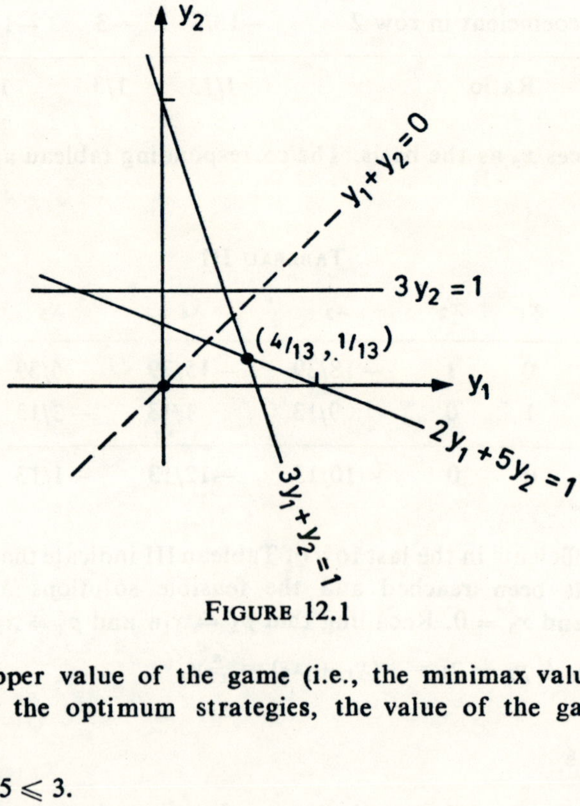

FIGURE 12.1

2 and the upper value of the game (i.e., the minimax value) is 3; and according to the optimum strategies, the value of the game is 13/5. Hence,

$$2 \leqslant 13/5 \leqslant 3.$$

Problem 12.2 For the 3 × 3 matrix game shown in Table 12.3, find the optimal strategy.

We make all the entries in the matrix positive by adding 10 to each element. This does not change the optimal strategy, but the value of the game is increased by 10, as shown in Table 12.4.

TABLE 12.3

		Player B 1	2	3
	1	4	−6	8
Player A	2	−6	8	−10
	3	8	−10	12

TABLE 12.4

		Player B 1	2	3
	1	14	4	18
Player A	2	4	18	0
	3	18	0	22

The problem now is:

Minimize $Z = x_1 + x_2 + x_3$

subject to

$$14x_1 + 4x_2 + 18x_3 \geqslant 1,$$
$$4x_1 + 18x_2 \geqslant 1,$$
$$18x_1 + 22x_3 \geqslant 1,$$
$$x_1, x_2, x_3 \geqslant 0.$$

Introducing slack variables, we obtain

$$14x_1 + 4x_2 + 18x_3 - x_4 = 1,$$
$$4x_1 + 18x_2 - x_5 = 1,$$
$$18x_1 + 22x_3 - x_6 = 1,$$
$$x_1, x_2, \ldots, x_6 \geqslant 0.$$

Tableau I gives the initial solution.

TABLEAU I

Basis	x_1	x_2	x_3	x_4	x_5	x_6	
x_4	−14	−4	−18	1			−1
x_5	−4	−18			1		−1
x_6	−18		−22			1	−1
	−1	−1	−1				0

To determine the new basis, we derive the ratios as in Table 1.

RATIO TABLE 1

non-basic variable	x_1	x_2	x_3
coefficient in last row	−1	−1	−1
coefficient in row 1	−14	−4	−18
Ratio	1/14	1/4	***1/18***

Hence, x_3 replaces x_4 as the basis in Tableau II.

TABLEAU II

Basis	x_1	x_2	x_3	x_4	x_5	x_6	
x_3	7/9	2/9	1	−1/18	0	0	1/18
x_5	−4	−18	0	0	1	0	−1
x_6	−8/9	44/9	0	−11/9	0	1	2/9
	−2/9	−7/9	0	−1/18			1/18

The corresponding ratios are as in Table 2.

RATIO TABLE 2

non-basic variable	x_1	x_2	x_3
coefficient in last row	−2/9	−7/9	−1/18
coefficient in row 2	−4	−18	0
Ratio	1/18	***7/162***	

Choosing the minimum ratio, we have x_2 replacing x_5 as the basis in Tableau III.

TABLEAU III

Basis	x_1	x_2	x_3	x_4	x_5	x_6	
x_3	59/81	0	1			0	7/162
x_2	2/9	1	0			0	1/18
x_6	−160/81	0	0			1	−4/81
	−4/81	0	0	−1/18	−7/162		8/81

Again, forming the ratio table, we see that x_1 replaces x_6 as the basis in Tableau IV.

TABLEAU IV

Basis	x_1	x_2	x_3	x_4	x_5	x_6	
x_3	0	0	1				1/40
x_2	0	1	0				1/20
x_1	1	0	0				1/40
	0	0	0				1/90

Tableau IV displays the optimal solution, with

$$x_1 = 1/40, \qquad x_2 = 1/20, \qquad x_3 = 1/40.$$

Recalling that $p_i = x_i w$, we have

$$1 = w(\tfrac{1}{40} + \tfrac{1}{20} + \tfrac{1}{40})$$

or

$$w = 10,$$

and the optimal strategy for A is (1/4, 1/2, 1/4). Remembering that the original pay-off matrix was increased by 10, we get the value of the original game as zero.

If the pay-off matrix consists of fractional elements, we can multiply all the elements of the matrix by a suitable factor k, thus making the elements integers. This does not alter the optimal strategies of the players, but the value of the game gets multiplied by k.

EXERCISES

1. For the pay-off matrices (a) and (b), determine the optimal strategies for players A and B and the value of each game.

(a)

	B		
A	6	2	7
	1	9	3
	−2	3	5

(b)

	B	
A	7	2
	5	5

[*Ans.* (a) A: $(\frac{7}{12}, \frac{5}{12}, 0)$; B: $(\frac{2}{3}, \frac{1}{3}, 0)$; value: $\frac{13}{3}$; (b) A: (0, 1); B: (0, 1) or $(\frac{3}{5}, \frac{2}{5})$; value: 5.]

2. Two players A and B choose a number from

00, 01, 10, 11

If the second digit of A's number coincides with the first digit of B's number, then A gets a pay-off equal to the sum of the first digits of A's and B's selections. A similar rule applies to B's pay-off. Write down the pay-off matrix for A and determine the optimal strategy and the value of the game.

[*Ans.*

	00	01	10	11
00	0	0	−1	0
01	0	0	0	1
10	1	0	0	−2
11	0	−1	2	0

optimal strategy is pure strategy: 01; value: 0.]

3. Transform the matrix games (a) and (b) into their corresponding primal and dual linear programming problems. Determine the optimal strategies for both the games and also their respective values.

(a)

2	1	0	−2
1	0	3	2

(b)

0	−1	1
1	1	−1
1	−1	0

4. Write the associated game matrix pay-off for the linear programming problem:

$$\text{Maximize } W = y_1 + y_2 + y_3$$

subject to the constraints

$$y_1 + 4y_2 + 3y_3 \leqslant 1,$$

$$4y_1 - 4y_2 + 2y_3 \leqslant 1,$$

$$3y_1 - y_2 + 5y_3 \leqslant 1,$$

$$y_1, y_2, y_3 \geqslant 0.$$

5. For the pay-off matrix shown here, determine the optimal strategies for players A and B and the value of the game:

		B		
		1	2	3
	1	1	−1	3
A	2	3	5	−3
	3	6	2	−2

[*Ans.* A: ($\frac{2}{3}$, $\frac{1}{3}$, 0); B: (0, $\frac{1}{2}$, $\frac{1}{2}$); value: 1.]

6. For the pay-off matrix given here, determine the optimal strategies for A and B and the value of the game:

		B			
		1	2	3	4
	1	3	2	4	0
A	2	3	4	2	4
	3	4	2	4	0
	4	0	4	0	8

[*Ans.* A: (0, 0, $\frac{2}{3}$, $\frac{1}{3}$); B: (0, 0, $\frac{2}{3}$, $\frac{1}{3}$); value: $\frac{8}{3}$.]

Index